ACPL ITEM
DISCARDED

SOLAR CENSUS

Photovoltaics Edition

aatec publications, p.o. box 7119, ann arbor, michigan 48107

To Richard and Joel

Allen County Public Library
Ft. Wayne, Indiana

Copyright © 1984 by **aatec publications**
P.O. Box 7119, Ann Arbor, Michigan 48107

Library of Congress Catalog Card No. 84-71602
ISBN 0-937948-05-5

All Rights Reserved
Manufactured in the United States of America

Cover design by Carl Benkert

CONTENTS

Directory Listings 1

Alphabetical listing of organizations—manufacturers, suppliers, designers, R&D, education and information sources—active in photovoltaics. Compiled from PV questionnaires, each entry contains current address, contact person/phone number, and activity description reflecting the evolution of this important technology.

Contact Name Index...............................163

Alphabetical listing of names of individuals affiliated with photovoltaics organizations.

Geographical Index................................175

Alphabetical listing of photovoltaics firms, organized state by state and internationally.

Subject Index......................................195

Topical breakdown of products, technological specializations, services, media, and research areas.

Every attempt has been made to provide a comprehensive, up-to-date directory. Because the photovoltaics field continues to change, with organizations entering and leaving the market, listings can become dated rather rapidly.

Neither the inclusion of those listed or the exclusion of others should be considered approval or disapproval by the publisher.

—**aatec publications**

A

ABACUS CONTROLS INC.
P.O. Box 893
Somerville, New Jersey 08876

201 526 6010

Manufacturer of inverters, converters, UPS.

ACHEVAL WIND ELECTRONICS
361 Aiken Street
Lowell, Massachusetts 01854

Don Bingley
617 453 0874

Manufacturer of power conditioning equipment. Line interconnect inverters.

ACP SOLAR POTENTIALS
1349 Capitol Street, N.E.
Salem, Oregon 97303

Dan Trumbull
503 363 1022

Supplier of PV components and systems. Technical assistance and economic analysis.

ACRO ENERGY
2006 S. Baker Avenue
Ontario, California 91761

714 947 4100

Constructs PV concentrating systems for commercial and utility applications, in conjunction with Applied Solar Energy Corporation.

ACUREX SOLAR CORPORATION
485 Clyde Avenue
Mountain View, California 94042

J. Schaeffer, Engineering Manager
415 964 3200

Engineering design of large PV systems for utility-scale electric power generation.

ADB ENGINEERS, INC.
641 East Beach Boulevard
Dania, Florida 33004

Arthur D. Benjamin, P.E., President
305 920 0200

Consulting engineers with extensive experience in solar energy systems. Our services include application feasibility studies and reports, conceptual and final design.

ADVANCED ENERGY CORP.
14933 Calvert Street
Van Nuys, California 91411

213 782 2191

Photovoltaic inverters.

ADVANCED SOLAR PRODUCTS, INC.
370 South Dixie Highway
Coral Gables, Florida 33133

305 666 5119

Distributor and retailer of PV panels and total systems, including the sale of refrigerators, lights, pumps and batteries.

AFG INDUSTRIES, INC.
P.O. Box 929
Kingsport, Tennessee 37662

John Helms
615 229 7200

Manufacturer of Solatex 2, fully tempered low-iron smooth-finish glass for use in PV panel covers; has high transmissivity at acute angles of incidence.

A.H.S. ENERGY SUPPLY
5547 Central
Boulder, Colorado 80301

Dan Cannon
303 449 0111

A.H.S. is currently involved with photovoltaics used to power water pumping systems on ranches in Colorado. We have designed a PV/solar-thermal system which is used to prevent freezing in remote cattle water troughs. All these units are fully portable and require no auxiliary energy.

AIRBORNE SALES
P.O. Box 2727
Culver City, California 90230

Catalog of surplus equipment, including 12-volt DC tools, etc.

AIR CONDITIONING AND REFRIGERATION INSTITUTE
1815 N. Fort Meyers Drive
Arlington, Virginia 22209

703 524 8800

AIRTRICITY
11145 Tampa Avenue
Suite 19-B
Northridge, California 91326

Charles Whitlock, President
818 368 1951

Wind farm developer, in joint venture with Holec to incorporate photovoltaics into their wind farm systems. Completed a 360-acre wind park in Tehachapi, California, with a 9-mW capacity.

ALDER/BARBOUR MARINE SYSTEMS INC.
43 Lawton Street
New Rochelle, New York 10801

Manufacturer of 12-volt DC refrigeration systems.

ALDERMASTON SALES
P.O. Box 34
Locust Valley, New York 11560

Malcolm Bru
516 676 6198

Manufacturer of PV-powered gift items—music boxes, oilwells, radios, calculators, etc.

ALPHA SOLARCO
1014 Vine Street, Suite 2530
Cincinnati, Ohio 45202

513 621 1243

Complete PV type solar systems for commercial, industrial and residential production of electric power.

ALTEK SYSTEMS, INC.
325 S. Union Street
Aurora, Illinois 60505

312 898 7000

Manufacturer of battery cabling systems.

ALTERNATE ENERGY TRANSPORTATION NEWSLETTER
c/o Electric Vehicles Consultants
327 Central Park West
New York, New York 10025

Ed Campbell
212 222 0160

Information on electric vehicles.

ALTERNATIVE ENERGY ENGINEERING, INC.
P.O. Box 339
Redway, California 95489

David Katz/Roger Herick
707 923 2277 or 707 923 3962

Manufacturer of auxiliary PV equipment: charge controllers, hydro systems, battery chargers, fuse boxes. Also, supplier of both PV components, auxiliary equipment, DC-powered equipment, DC schematics, how-to diagrams. 50-page catalog available.

ALTERNATIVE SOURCES OF ENERGY
107 South Central
Milaca, Minnesota 56353

Abby Marier
612 983 6892

Bi-monthly magazine. Covers all phases of alternative energy; PV section in each issue.

ALTERNOLOGIES
Box 1008
Fort Collins, Colorado 80522

303 223 5395

Microcomputer software for residential/commercial designers in the renewable energy field.

AMERICAN ENERGY CONSULTANTS
8444 Melba Avenue
Canoga Park, California 91304

213 346 7240

Microcomputer software for residential/commercial designers in the renewable energy field.

AMERICAN INSTITUTE OF ARCHITECTS
1735 New York Avenue, N.W.
Washington, D.C. 20006

202 626 7500

AMERICAN POWER CONVERSION CORPORATION
89 Cambridge Street
Burlington, Massachusetts 01803

Ervin F. Lyon, President
617 273 1570

Our product line for residential photovoltaic systems encompasses all the major components, exclusive of PV array, necessary to implement a utility-interactive solar electric system. This includes the DC-to-AC inverter, the Photovoltaic Source Combiner Unit, and the Lightning Protection/Grounding Unit. Thus the installer need only add the wiring, ground rod, conduit and associated fittings to complete the system. We provide comprehensive system design and application information to assist the builder, developer, installer or homeowner in the implementation of a PV system. Research activities include PV power conversion.

AMERICAN SOCIETY OF HEATING, REFRIGERATING
AND AIR-CONDITIONING ENGINEERS
1791 Tullie Circle, N.E.
Atlanta, Georgia 30329

404 636 8400

AMERICAN SOCIETY OF MECHANICAL ENGINEERS
345 W. 47th Street
New York, New York 10017

212 705 7722

AMERICAN SOLAR ENERGY SOCIETY, INC.
1230 Grandview
Boulder, Colorado 80302

303 492 6017

AMERICAN STANDARDS TESTING BUREAU INC.
40 Water Street
New York, New York 10004-2672

Robert Cook
212 943 3160

Complete testing laboratory facilities for product failure and/or product certification.

AMERICAN SUN
7541 Lemp Avenue
North Hollywood, California 91605

Jack R. Waizenegger
213 765 2184

American Sun specializes in the design and marketing of photovoltaic power systems for remote homes, water pumping, communications, navigational aids, lighting, cathodic protection, transportation, data transmission, recreational vehicles, boats, and field offices. A full line of photovoltaic modules, support structures, batteries, controllers, inverters, and accessories are available wholesale and retail.

AMERICAN WIND ENERGY ASSOCIATION
1516 King Street
Alexandria, Virginia 22341

703 684 5196

AMETEK INC.
Station Square Two
Paoli, Pennsylvania 19301

215 647 2121

R&D on cadmium and cadmium telluride thin-film photovoltaic cells. Commercial production of Czochralski single-crystal silicon wafers.

AMFRIDGE CORPORATION
23892 Cooper Drive
Elkhart, Indiana 46514

219 262 2521

Manufacturer of DC refrigeration equipment.

AMP INCORPORATED
Harrisburg, Pennsylvania 17105

Tom Sotolongo, Senior Development Engineer
717 564 0100

Manufacturer of the SOLARLOK inter-module connector for PV systems. The connectors feature a high-pressure contact system with a 30-ampere rating, a quick disconnect feature, and an environment-resistant thermoplastic housing.

APPLIED SOLAR ENERGY CORP.
15251 E. Don Julian Road
City of Industry, California 91746

B. Thompson, President
213 968 6581

R&D silicon cells, photosensors. Solar cells and panels for space and terrestrial applications.

APPROPRIATE DESIGNS & CONSTRUCTION
Box 397A Steuben Valley Road
Holland Patent, New York 13354

John Siegenthaler
315 865 8903

Appropriate Designs & Construction is a consulting firm working originally in the areas of active and passive solar design/construction. The firm now has capabilities to provide state-of-the-art systems analysis and design for stand-alone and utility-interfaced photovoltaic electrical systems. Residential and remote site equipment power supplies may be evaluated and specified.

Appropriate Designs is currently conducting research on PV module/DC pump combinations for use in powering residential-scale solar domestic hot water systems. PV module/DC pump combinations are laboratory tested. Empirical correlations of the test data are then used in conjunction with an hourly simulation computer program to predict daily and seasonal performance of the specified system.

APPROPRIATE ENERGY MANAGEMENT
10125 Lilly Chapel
Georgesville Road
West Jefferson, Ohio 43162

Wes Eggleton
614 879 7481

Supplier of PV and auxiliary equipment. Also, design and construction of custom DC power units; design and manufacture of hybrid wind/PV systems. Registered P.E. electrical engineer on staff with PV manufacturing and design experience, DC power systems design, and wind installation design.

ARCO SOLAR, INC.
P.O. Box 4400
Woodland Hills, California 91365

Cedric Grgurich, Marketing Communications
213 700 7000

ARCO Solar, Inc. manufactures photovoltaic modules using single-crystal silicon cells in a variety of electrical outputs. The company also offers support structures and accessory components for complete DC and AC power systems from small single-module systems to multi-megawatt utility-scale systems. Throughout the world, ARCO's modules are powering telecommunications stations, water delivery systems, homes and cabins, navigational aids and railroad signals, utility-scale power plants and more.

ARCO Solar markets its products and systems worldwide through a network of regional offices, trained distributors and their dealers, who aid and advise on the design, installation and maintenance of both DC and AC photovoltaic systems. Its manufacturing and design capabilities are backed by the largest private photovoltaic research program (thin-film) in the world.

ARCO Solar, a wholly owned subsidiary of Atlantic Richfield Company, owns and operates, with Southern California Edison metering and purchasing the electricity, the one-megawatt Hesperia Power Plant and the 4.5-megawatt Carrisa Plain Power Plant. ARCO Solar publishes the *ARCO Solar News*.

ARCO SOLAR FAR EAST PTY. LTD. — NORTH
Toranomon 37 Mori Building
5-1 Toranomon 3-chome
Minato-ku Tokyo 105
Japan

Tom Dyer
03 434 6191

Regional office of ARCO Solar, Inc.

ARCTIC-KOLD ENERGY PRODUCTS
Division, Birken Manufacturing Company
3 Old Windsor Road
Bloomfield, Connecticut 06002

David Gurne, Marketing Director
203 242 2211

Arctic-Kold manufactures DC-, solar- and wind-powered refrigerators and freezers for marine, mobile and remote-site use. Products include 5.3 and 8.1 cubic feet free-standing refrigerator/freezers and ice box conversion kits, which will cool an insulated enclosure with a volume area up to 10 cubic feet—all utilizing Arctic-Kold's refrigerator compressor, driven by a 1/8-hp 1800-rpm permanent magnet motor. The motor is coupled to the compressor by a magnetic clutch, eliminating the need for a driveshaft seal and is closed with socket cap screws rather than welded shut, thus easily serviced or repaired. Electrically efficient, the compressor requires only 60 amps per day. Acceptable sources of DC electricity include batteries, gasoline, water and wind generators, and PV panels.

Also available are single-stage submersible DC micro-irrigation pumps. Soon to be available from Arctic-Kold is a solar-powered medical refrigerator which requires no batteries. Research activities include the development of a brushless DC motor-driven water pump for deep wells and micro-irrigation.

AREMCO PRODUCTS, INC.
P.O. Box 429
Ossining, New York 10562

G. Lawrence Grimaldi, Product Manager
914 762 0685

AREMCO produces silicon cell manufacturing equipment: silicon wafer screen printing equipment (thick-film), AREMCO Accu-Coat

Vidalign Model 3232 and diamond dicing saws (silicon), Accu-Cut models 5200 and 5250.

UNIVERSITY OF ARIZONA
Solar and Energy Research Facility (SERF)
College of Engineering
Tucson, Arizona 85721

Dr. Rocco Fazzolare, Director
602 626 0184

Research includes segmented parabolic PV collectors, monitoring programs, data analysis, commercial/industrial applications of photovoltaic power.

ARIZONA PUBLIC SERVICE CO. (APS)
General Offices
411 N. Central Avenue
Phoenix, Arizona 85004

602 271 7900

The Sky Harbor 225-kWp concentrating photovoltaic facility is intertied with APS. Designed by Martin-Marietta, activated April 17, 1982.

ARIZONA STATE UNIVERSITY
Center for Solid State Science (CSSS)
Tempe, Arizona 85281

J. Bruce Wagner, Director
602 965 4544

Research on photovoltaic materials.

ARKWORK REVIEW
Denton County Arkwork
711 West Sycamore
Denton, Texas 76201

Tom Miller, Editor
817 387 1659

Arkwork provides information, experimental models and practical demonstrations of appropriate technology. Publishes the *Arkwork Review*, a quarterly journal.

ASA GOVERNMENT MARKETING SERVICE
1127 So. Patrick Drive, Suite 15
Satellite Beach, Florida 32937

Al Stern, President
305 777 2018

Marketing for sales to government agencies and industry. We find opportunities for companies to sell energy products to the government. We also find opportunities for them to bid on government contracts that provide research and development funds for engineering of energy systems. We also assist companies in their proposal writing necessary to win the government contract awards.

ASSOCIATES IN RURAL DEVELOPMENT, INC.
P.O. Box 1397
Burlington, Vermont 05402

Rick McGowan, Senior Engineer
802 658 3890

ARD is working with PV-driven water pumps, refrigeration units, passive solar monitoring equipment, and anemometry equipment all associated with its USAID-sponsored project, the Botswana Renewable Energy Technology Project, in southern Africa. We are setting up a comparative testing program for five PV pumping systems spread throughout the country.

An upcoming project will compare a biomass gasifier powered water pump with a low-head PV pump, in Malaysia. A variety of PV applications will be included in an energy project in Barbados and Antigua which will be starting summer 1984.

ASSOCIATION OF ENERGY ENGINEERS
4025 Pleasantdale Road, Suite 340
Atlanta, Georgia 30340

Albert Thurmann
404 447 5083

Membership association.

ATLANTIC SOLAR POWER, INC.
6455 Washington Boulevard
Baltimore, Maryland 21227

Brent R. Atkins, General Manager/Doug Keller
301 796 8000

ARCO Solar distributor. Distributes balance of systems including supports, batteries, control housings, modules and groundings to dealers and the OEM network throughout the U.S. and overseas. Also provides system design, analysis and fabrication.

ATR ELECTRONICS, INC.
302 E. Fourth Street
St. Paul, Minnesota 55101

612 222 3791

Manufacturer of inverters.

B

BACKWOODS CABIN ELECTRIC SYSTEMS
8530 Rapid Lightning Creek Road
Sandpoint, Idaho 83864

Steve Willey
208 263 4290

Manufacture meterbox and Pelton power unit. Distributor for ARCO Solar. Supply components for PV, wind and hydro systems. Customize stereo, telephone and radio setups for remote sites. Teach classes in alternative energy and conduct seminars and workshops.

BALANCE OF SYSTEMS SPECIALISTS, INC. (BOSS)
7745 E. Redfield Road
Scottsdale, Arizona 85260

Bradley E. O'Mara, President
602 948 9809

BOSS, Inc. researches, designs, develops, manufactures and supplies electronic products for the PV industry, foreign and domestic. Products include 12-volt battery voltage regulators; LED light bar annunciators to indicate system performance and hook-up errors; battery charging controllers (up to 120 volts); power converters; cathodic protection control; PV pool pumping systems.

Provides engineering services to PV industry distributors and dealers, including computer-aided systems sizing.

BOSS, with Photocomm, recently acquired the manufacturing and supply capabilities of Photowatt International.

BALL STATE UNIVERSITY
Center for Energy Research/Education/Service (CERES)
Muncie, Indiana 47306

Robert J. Koester, Director
317 285 4938

Primary emphasis of CERES research—ranging in scale from community planning to materials technology—is the interrelation of disciplines in energy-related methods of analysis, performance evaluation and decision-making, and deriving new knowledge bases by means of physical experiments, computer modeling/simulations and case studies.

CERES is housed in a 15,000 square foot laboratory and classroom building containing full-scale demonstrations of passive and active solar technology. A broad framework of educational activities is available to students at Ball State and through continuing education classes both on and off campus. Available for the Indiana community are on-site workshops and energy awareness programs.

BALL STATE UNIVERSITY
Department of Physics and Astronomy
Muncie, Indiana 47306

Ronald M. Cosby, Professor
317 285 6268

Research: Analysis of concentrating photovoltaic systems, particularly line-focusing Fresnel lens – solar cell systems. Involves modeling of both the concentrator and the photovoltaic cell.

Instruction: Introduction to Solar Energy Technology & Applications, undergraduate, low mathematical level course. Applied Physics, advanced course in physics, content primarily photovoltaics.

BARRETT HEATING & AIR CONDITIONING CO., INC.
2260 Union Boulevard
Bayshore, New York 11706

Gary L. Shoemaker, P.E.
516 665 0940

HVAC/solar contractor. Design, engineer, sell and install HVAC/solar/PV systems.

BATTELLE COLUMBUS LABORATORIES
505 King Avenue
Columbus, Ohio 43201

614 424 6424

Solar cell research and development.

BERKELEY SOLAR GROUP
3140 Grove Street
Berkeley, California 94703

415 843 7600

Microcomputer software for residential/commercial designers in the renewable energy field.

BERTOIA STUDIO, LTD.
P.O. Box 383
Bally, Pennsylvania 19503

Val Bertoia
215 845 7096

Bertoia Studio, Ltd. is involved in manufacturing wind-electric systems. Mr. Bertoia uses photovoltaics in his home wind system. He has also designed a stone-walled greenhouse to protect the photovoltaics system from adverse weather conditions. Bertoia Studio Ltd. also sells books on renewable energy.

BEST ENERGY SYSTEMS, INC.
P.O. Box 280
Necedah, Wisconsin 54646

Marguerite M. Paul, Vice President
608 565 7200 or 1 800 356 5794

Manufactures 22 models of inverters (from 1000 to 12,000 watts), plus accessories and options. Also, designs and manufactures power protection systems (outage protection systems, OPS) and mobile power systems. Distributes "How to Design an Independent Power System," by Best president, Terrance D. Paul.

BIOMASS ENERGY RESEARCH ASSOCIATION
1377 K Street, N.W.
Washington, D.C. 20005

202 833 3405

BLACK & VEATCH, ENGINEERS – ARCHITECTS
1500 Meadow Lake Parkway
P.O. Box 8405
Kansas City, Missouri 64114

J. Charles Grosskreutz/Sheldon L. Levy
913 967 2554 or 913 967 7119

B&V's orientation is primarily large systems. We serve the utility and commercial/industrial firms in the private sector and all levels of government in the public sector. Experience in: engineering design of PV systems—includes electrical, structural and safety analysis; PV systems performance analysis; research and development of PV systems; and design of concentrating PV arrays.

BLACK HAWK ASSOCIATES, INC.
243 E. 19th Avenue, Suite 310
Denver, Colorado 80203

David Schaller
303 861 0301

Black Hawk Associates provides consulting services to communities, homeowners, potential agricultural and small industrial users, and local governments in the analysis and implementation of photovoltaic energy systems. Capabilities of the firm include solar data monitoring and analysis, energy use definitions, site evaluations, technology assessment, system sizing and cost analyses, institutional studies, and project management.

Black Hawk Associates is qualified to conduct photovoltaic system analyses, particularly for remote, stand-alone applications. Black Hawk Associates has recently conducted a U.S. Department of Energy-sponsored photovoltaic system sizing and cost analysis for seven governments in the U.S. Pacific territories. This analysis selected several potential end-use applications, including village power, water pumping, and refrigeration for sizing, performance and cost evaluation.

BLISS MARINE
Route 128 at Exit 61
Dedham, Massachusetts 02026

Marine equipment catalog carrying 12-volt DC items.

BOEING ENGINEERING AND CONSTRUCTION COMPANY
P.O. Box 3707
Seattle, Washington 98124

1 800 243 8160

Recently joined Reading & Bates Development Co. of Tulsa, Oklahoma, in forming Solvolco to develop thin-film solar cell modules for commercial markets. Continuing R&D in improvement of polycrystalline PV cells.

BORREGO SOLAR SYSTEMS
The Center, Suite 301
590 Palm Canyon Drive
Borrego Springs, California 92004

Residential design services. Distributor for ARCO Solar, Inc.

BRADEN WIRE AND METAL PRODUCTS
P.O. Box 5087
San Antonio, Texas 78201

512 734 5189

PV automatic feeders, gate openers.

BROKEN PLOW LAW OFFICE
Star Route 1, Box 33B
Chadron, Nebraska 69337

Andrew B. Reid/Phyllis Girouard
308 432 4259

Environmental law practice, housed in a photovoltaic-powered office.

BROOK FARM INC.
Brook Road
Falmouth, Maine 04105

David Sleeper
207 797 9380

Designs and constructs electrical systems for PV homes. Supplier of PV and auxiliary equipment.

BROOKHAVEN NATIONAL LABORATORY
Upton
Long Island, New York 11973

516 282 2123

Research and development, photovoltaic conversion.

C

CALIFORNIA ENERGY COMMISSION
1516 Ninth Street
Sacramento, California 95814

Arthur J. Soinski
916 324 3467

Technology assessment, market assessment, demonstration projects, technical assistance.

CAMBRIDGE SOLAR ENTERPRISES, INC.
Cambridge Photovoltaics Industry, Research Division
55 Wheeler Street
Cambridge, Massachusetts 02138

Dennis Reinhardt, President
617 354 5272

Dealer for ARCO. Residential and marine systems, including sizing and installation services. Agent for Spire Corporation, setting up joint ventures in West Africa, India, Malaysia and Singapore, for fabrication of modules. R&D in PV end applications: low-power telecommunications and pumps.

CAM-LOK DIVISION
Empire Products, Inc.
10540 Chester Road
P.O. Box 15888
Cincinnati, Ohio 45215

Kevin Davis
513 771 3171

Manufacturer of electrical connectors. CAM-LOK "J" series connectors—for use on photovoltaic arrays or wind turbine generators. They are rugged, waterproof and have low voltage drop characteristics, and can be connected and disconnected without tools. Also design connectors per customer specifications.

CANADIAN SOLAR INDUSTRIES ASSOCIATION
67A Sparks Street
Ottawa, Ontario K1P 5A5
Canada

CARBONE INVESTMENT MANAGEMENT CORPORATION
2000 Center Street, No. 1291
Berkeley, California 94704

Robert C. Carbone
415 843 2835

CIMCO is a registered investment advisor specializing in the photovoltaics industry. Three PV divisions: (1) Publishes *Photovoltaic Investment Newsletter*, a monthly investment analysis and research of the photovoltaic industry; conducts PV investment seminars. (2) Investment management services for institutions and qualified individuals seeking investment opportunities in the PV industry; founder of Photovoltaic Investment Fund. (3) Financial consulting services include finding capital for PV companies; assist in developing third-party and private offering memorandums for PV companies; facilitate PV mergers, acquisitions and joint ventures; and locate PV projects.

CARTER WIND SYSTEMS, INC.
Route 1, Box 405-A
Burkburnett, Texas 76354

817 569 2238

Variable voltage, high-efficiency generators. Microprocessor performance monitoring.

C&D POWER SYSTEMS
3043 Walton Road
Plymouth Meeting, Pennsylvania 19462

Kenneth Krodt
215 828 9000

Manufactures batteries specifically designed for use with photovoltaic systems, both shallow-cycle (can be discharged up to 20% of capacity daily) and deep-cycle (can be discharged up to 80% daily). Standard available options include: enclosures for shallow-cycle batteries; gas recombiners for both shallow- and deep-cycle batteries; and interconnects and electrolyte circulation pumps for deep-cycle batteries.

C&D BATTERIES
150 Connie Crescent
Concord, Ontario L4K 1B6
Canada

CHARLINS, INC.
P.O. Box 2237
Hudson, Ohio 44236

B. Charlins, President
216 656 3431

Manufacturer of syphonic pump systems.

CHRONAR CORPORATION
P.O. Box 177
Princeton, New Jersey 08542

George Self, Director of Marketing
609 587 8000

Chronar Corporation, a research and manufacturing concern, manufactures amorphous silicon PV panels. They also provide technical services, as well as custom design and installation of systems to specification. Products include the CPV-4030, a 30-cell amorphous silicon PV panel; the 1/3-watt Pocket Charger, to charge two 1.2-volt ni-cad batteries; the Power Pack, a charger and storage device that provides standby or direct power to 12-volt devices; the CPV-1001, a 5-cell PV power source for LCD calculators, instruments, etc; and Solar Roof Shingles. Research activities include increased efficiency single-junction, thin-film amorphous silicon cells. In conjunction with Chronar, Thomas & Betts is developing a PV Raceway Interconnect System. With AFG Industries, they are planning to build an amorphous solar cell manufacturing plant in Port Jervis, New York.

CITIZENS ENERGY PROJECT
1110 Sixth Street, N.W., Suite 300
Washington, D.C. 20001

202 289 4999

COLORADO MOUNTAIN COLLEGE
3000 County Road 114
Glenwood Springs, Colorado 81601

Steve McCarney, Associate Professor
303 945 7481

Offers solar education courses, including PV.

COLORADO TECHNICAL COLLEGE
655 Elkton Drive
Colorado Springs, Colorado 80907

Donald D. Mueller
303 598 0200

Photovoltaics course—a study of the photovoltaic generation of electrical power to include design and sizing of a small stand-alone system, theory of operation of a PV cell, concentrator systems, cell and module production, emerging materials, and present and future applications.

COMMUNICATIONS ASSOCIATES
305 N. Republic Avenue
Joliet, Illinois 60434

James Hartley, Photovoltaic Products Manager
815 744 6444 or 1 800 435 9313

Distributor for ARCO Solar and PV balance of system component manufacturers. Systems design and engineering. Dealer network.

COMPUSOLAR
Gum Springs Road
Jasper, Arkansas 72641

Stephen Cook
501 446 2211

Microcomputer programs for photovoltaic system design. Also residential solar electric system slide programs. Provide consulting services for PV projects, specializing in residential and remote applications.

COMPUTER POWER INC.
124 W. Main Street
High Bridge, New Jersey 08829

201 735 8000

Manufacturer UPS, IPS, inverters, battery chargers, and other electronic components.

COMPUTER SHARING SERVICES, INC.
7535 E. Hampden Avenue, Suite 200
Denver, Colorado 80231

303 695 1500

Microcomputer software for residential/commercial designers in the renewable energy field.

CONDAR CO.
12000 Winrock Road
Hiram, Ohio 44234

Peter Cornelison, Vice President
216 569 3245

Manufacturer of the Solarmeter window thermometer and the Pipe Temp pipe thermometer. The Solarmeter reads Btu/sq ft/hr from 150 to 400; also shows temperatures reached on a black plate collector. Attaches instantly to window with suction cup. Pipe Temp snaps on pipes and records fluid temperature from 50 to 250°F; 1-inch-diameter dial.

CONSERVATION AND RENEWABLE ENERGY INQUIRY AND REFERRAL SERVICE (CAREIRS)
P.O. Box 1607
Rockville, Maryland 20850

1 800 523 2929

UNIVERSITY OF CONNECTICUT
Solar Energy Test Laboratory
Box U-139
Engineering Building II, Room 306
Storrs, Connecticut 06268

T. A. MacQueen, Jr., Director
703 486 3478

Photovoltaics, research and development.

COPLEY ENERGY, INC.
2507 Lisbon Lane
Alexandria, Virginia 22306

Ernest L. Copley, III, President

Owner and operator of the PV-powered Copley Energy Plant, intertied to Delmarva Power & Light Co. The 15.7-kWp plant was designed by Mueller Associates of Baltimore, Maryland.

COSMOS DEVELOPING ASSOCIATES, INC.
P.O. Box 3808
Vero Beach, Florida 32964

Terry Torres
305 231 7977

Supplies cells, arrays, modules, batteries, components and PV-powered equipment. Consultant and system design services.

CREATIVE ELECTRONICS
221 N. Lasalle Street, Suite 1038
Chicago, Illinois 60601

Inverter manufacturer.

CRYSTAL SYSTEMS, INC.
35 Congress Street
Salem, Massachusetts 01970

Clifford J. Herman, Sales Manager
617 745 0088

We manufacture 32 x 32 cm square silicon ingots for photovoltaic and optical applications.

CURRENT ALTERNATIVES
College Town Industrial Plaza
P.O. Box 166
Northfield, Vermont 05663

Doug Pennington
802 485 6952

Current Alternatives is sole distributor in the northeast United States for Solec International photovoltaic products. We design and market complete systems for industrial and commercial applications in addition to the private sector.

CW ELECTRONIC SALES CO.
800 Lincoln Street
Denver, Colorado 80203

John Capone/Phil Leavenworth
303 832 1111

Distributor of PV products and accessories.

D

DALE & ASSOCIATES
2870 Bartells Drive
Beloit, Wisconsin 53511

608 362 1495

Manufacturer of 12-volt DC fans.

DANFOSS INC.
16 McKee Drive
Mahwah, New Jersey 07430

Steve M. Madigan
201 529 4900

Supplier of 180-Btu/hr R12 hermetic compressor with DC motor and electronic commutator/control device for solar-powered refrigerators. Design based on high-volume production AC compressor.

JOEL DAVIDSON, PV NETWORK NEWS
10615 Chandler Boulevard
North Hollywood, California 91601

213 980 6248

Founder of the PV Network, and the *PV Network News*, a solar electric consortium providing information for and from PV users. Originally intended for the remote home market, it has expanded to cover all aspects of photovoltaics. The newsletter contains an access and information section, systems design, letters, and equipment sales. Published quarterly. Mr. Davidson has developed, with Greg Johanson, a PV-powered 1-kW electric vehicle. PV consultant and supplier of PV equipment. Author of *The Solar Electric Home*.

DEFENSE PHOTOVOLTAIC PROGRAM OFFICE
U.S. Army Mobility Equipment R&D Command
Fort Belvoir, Virginia 22060

Photovoltaics R&D. Manages Mt. Laguna PV Air Force Station, California, one of the world's largest applications of solar cell equipment, completed in 1979.

DELAWARE DEPARTMENT OF ADMINISTRATIVE SERVICES
Division of Facilities Management/Energy Office
P.O. Box 1401
Dover, Delaware 19903

Edward M. Shimamoto
302 736 5644

Public information and outreach on alternative renewable energy resources.

UNIVERSITY OF DELAWARE
Institute of Energy Conversion
One Pike Creek Center
Wilmington, Delaware 19808

302 995 7155

Advanced PV cell research. Currently working on design of high-efficiency thin-film solar cell, attempting to enhance the 10% efficiency for copper indium diselenide.

DEPARTMENT OF ENERGY, U.S. (DOE)
Photovoltaics Energy Technology Division
1000 Independence Avenue
Washington, D.C. 20585

R. H. Annan, Director
202 252 5000

DOE has three regional photovoltaics test centers: Southwest Residential Experiment Staton (SW RES), Las Cruces, New Mexico, operated by New Mexico Solar Energy Center; Southeast RES, Cape Canaveral, Florida, a joint effort of Florida Solar Energy Center, Georgia Institute of Technology, and the Alabama Solar Energy Center; and the Northeast RES, Concord, Massachusetts, operated by MIT's Energy Lab.

DODGE PRODUCTS, INC.
P.O. Box 19781
Houston, Texas 77224

713 467 6262

Manufacturer of solar measuring instruments: net radiometers, pyranometers, pyroheliometers.

DOMETIC SALES CORP.
2320 Industrial Parkway
P.O. Box 490
Elkhart, Indiana 46514

219 294 2511

DC refrigeration equipment manufacturer.

MARK D. DOSTAL, ENGINEER
225 Vivian Lane
Stevens Point, Wisconsin 54481

715 344 2137

Active/passive solar system design and construction.

DREXEL UNIVERSITY
32nd and Chestnut Street
Philadelphia, Pennsylvania 19104

215 895 2000

Advanced R&D in copper indium diselenide polycrystalline thin-film solar cells.

DSET LABORATORIES INC.
Box 1850 Black Canyon Stage I
Phoenix, Arizona 85029

P. V. French, Customer Service Representative
602 465 7356

Photovoltaic testing laboratory. Tests on modules and cells include electrical performance, materials durability, and long-term module performance. Also, supplier of PV reference cells: assembly and calibration.

DUANE'S SOLAR ENERGY CO.
1625 Cottage Street, S.E.
Salem, Oregon 97302

Duane H. Bowen
503 362 9115

Duane's Solar Energy has been involved in photovoltaics for five years as a dealer and networker for Solarex. Also sells Best and Dytek products. Furnishes educational literature and information. PV catalog available.

DYNAMIC SOLAR PRODUCTS INC.
90270 Overseas Highway
Tavernier, Florida 33070

Robert T. Epstein
305 852 9683

Supplier of solar-powered aerator equipment for bait and fish tanks. Sailboat battery charging systems (house boat). RV equipment. Affiliated firms Alert 80 Energy Systems and Energy Systems Leasing provide energy management services.

DYNAMOTE CORPORATION
1200 W. Nickerson
Seattle, Washington 98119

206 282 1000

Manufacturer of MB Solid State Static Inverters, 120-volt 60 Hz AC power from batteries; 12, 24, 32, 36, 48 or 120 volts DC systems. SCR type modified square wave inverter has extra surge capacity for starting induction motors. Standard features: crystal-controlled oscillator; load demand start; low voltage cutout; 120/240 volts AC output. Options: 220 volts AC 50 Hz output for export; charger circuit; low voltage transfer circuit.

DYNA TECHNOLOGY INC.
7850 Metro Parkway
Minneapolis, Minnesota 55420

612 853 8400

Manufacturer of inverters, gen sets.

DYTEK LABORATORIES, INC.
40 Orville Drive
Bohemia, New York 11716

Mary Urso
516 567 8500

Manufacturer of high-performance inverters: ratings to 2500 watts continuous with up to 10,000-watt surge to start refrigerators, pumps and compressors. Efficiencies over 80% with high-quality quasi-sinewave output for operation of sensitive equipment. Automatic turn-on to conserve power. Available in 12, 24 and 32 volts DC input voltage.

E

THE EARTH STORE
P.O. Box 679
North San Juan, California 95960

Jon Hill
916 292 3395

The Earth Store sells alternative energy equipment—PV, wind and hydro, auxiliary supplies, and 12-volt appliances—through our retail store and through the mail. We also provide photovoltaic-powered electricity for sound systems at concerts and rallies. In business since 1976. Catalog available.

THE EARTH STORE
13224 Tyler Foote Road
Nevada City, California 95959

Jon Hill
916 292 3395

EASCO ALUMINUM
New Jersey Aluminum Solar Division
P.O. Box 73
North Brunswick, New Jersey 08902

M. L. Slom, Manager – Solar Division
201 249 6867

Manufactures, for both prototype and production, aluminum extrusion framings for PV module housings, absorbers, supports and mountings.

EATON CORPORATION
100 Erieview Plaza
Cleveland, Ohio 44114

216 523 5000

Advanced R&D in polycrystalline thin-film solar cells.

ECOTOPE, INC.
2812 E. Madison
Seattle, Washington 98112

206 322 3753

Microcomputer software for residential/commercial designers in the renewable energy field.

EDMONDS COMMUNITY COLLEGE
Energy Management Program
20000 – 68th Avenue, West
Lynnwood, Washington 98036

Terry Egnor/Gary Lintz
206 771 1693

The basic operation of PV cells and current research and application are taught in the context of a first-year course, Energy 140, Renewable Energy Systems.

EIC LABORATORIES
111 Chapel
Newton, Massachusetts 02158

617 965 2710

R&D—a nonvacuum alternative to solar cell fabrication.

ELECTRIC POWER RESEARCH INSTITUTE (EPRI)
P.O. Box 10412
Palo Alto, California 94303

Roger Taylor, PV Research
415 855 2000

Research and development of multijunction amorphous silicon, ribbon silicon, concentrating systems.

ELECTROLAB INC.
2103 Mannix
San Antonio, Texas 78217

Karl A. Senghaas
512 824 5364

Design and manufacture of PV-powered control systems for OEMs. Custom photovoltaic panels, solar-powered timers, feeders, ventilators, gate openers, actuators; solar-powered control systems for oil and water pumping.

ELECTRONIC DEVICES & CONTROLS INC.
1300 N.W. McNab Road
Ft. Lauderdale, Florida 33309

Temperature controls; PV systems and components.

ELTRON RESEARCH, INC.
710 E. Ogden Avenue
Naperville, Illinois 60540

312 369 6040

R&D—ways to predict the performance of semiconductor materials in solar cells.

ENCON PHOTOVOLTAICS
27600 Schoolcraft
Livonia, Michigan 48150

Pete DeNapoli
313 523 1850

Encon is a distributor of Solarex and ARCO Solar modules and other PV products. Manufacturer of small battery charging panels and balance of systems equipment. Power factor controllers. PV speaker and seminar services available.

ENERCOMP
2655 Portage Bay, Suite 6
Davis, California 95616

916 753 3400

Microcomputer software for residential/commercial designers in the renewable energy field.

ENERDYNE SOLAR & WOOD SYSTEMS
P.O. Box 366
Suttons Bay, Michigan 49682

Dick Cookman
616 271 6033

Supplier of PV equipment.

ENERGEIA
732 W. 6th Avenue
Eugene, Oregon 97402

Tom Scott
503 485 5719

Retail outlet for both residential and commercial PV users. Design services available.

ENERGY ALTERNATIVES
979-A East Avenue
Chico, California 95926

Chuck Alldrin
916 345 1722

Design/installation of PV systems. Supplier of modules, inverters, batteries, etc. Custom electronic control systems engineered to specifications.

ENERGY COMPLIANCE SYSTEMS, INC.
2881 Hemlock Avenue, Suite 2
San Jose, California 95128

David Horobin
408 244 1220

Our services in the photovoltaic area include: (1) solar design and consultation, and (2) solar energy system design and analysis.

ENERGY CONSERVATION & SOLAR CENTER
121 Valley Street
Manchester, New Hampshire 03301

603 625 9677

Full spectrum solar store, including photovoltaic components and accessories. 80-page catalog.

ENERGY CONVERSION DEVICES, INC.
1675 W. Maple Road
Troy, Michigan 48084

313 280 1900

R&D and production of Ovonic amorphous silicon cells (12-inch wide, continuous strip). In partnership with Standard Oil Co. (SOHIO) to commercialize ECD's Ovonics under the name Sovonics Solar Systems. The solar cell manufacturing plant is in Troy, Michigan, the panel factory in Warrensville, Ohio. Sharp-ECD manufactures Ovonic solar cells in Japan.

ENERGY EQUIPMENT SALES
412 Longfellow Boulevard
Lakeland, Florida 33801

Ron Vachabach
813 665 7085

Manufacturer's representative for numerous PV products, components and accessories.

ENERGY HARVESTER
13041 Roundup Avenue
San Diego, California 92128

619 485 8454

Concentrating collectors, PV power systems.

ENERGY MANAGEMENT ANALYSIS OF MADISON
314 S. Mills Street
Madison, Wisconsin 53715

608 255 1397

Microcomputer software for residential/commercial designers in the renewable energy field.

ENERGY MANAGEMENT CONSULTANTS INC.
672 S. Lafayette Park Place, No. 38
Los Angeles, California 90057

Douglas S. Stenhouse/James D. Roberts
213 383 3195

Multidisciplinary firm directed toward energy conservation design and analysis. Conventional disciplines of architecture, urban planning, mechanical and electrical engineering, plus environmental engineering, systems control design, solar energy system design, and computer programming. These capabilities result in innovative yet practical solutions to complex problems and large projects. Services include feasibility studies, preliminary and final design, as well as review of construction work and contractor submittals. Over the past 8 years, EMC has completed numerous solar energy feasibility studies through final design, including design/incorporation of photovoltaic systems.

ENERGY MATERIALS CORP. (EMC)
Sterling Road
South Lancaster, Massachusetts 01561

Jack Paster, Vice President
617 365 7383

R&D of ribbon and sheet growth cells—low angle silicon sheet (LASS), in partnership with Scicap.

ENERGY RESEARCH & DESIGN ASSOCIATES
35 S. Cache
Box 3177
Jackson, Wyoming 83001

Jim Kleyman
307 733 8018

Residential PV systems design. Do-it-yourself PV kit systems and workshops.

ENERGY REVIEW
2074 Alameda Padre Serra
Santa Barbara, California 93103

Susan Williams, Editor
805 965 5010

Energy Review is a bi-monthly publication that provides concise, readable digests of important articles, government and private reports, and recent books relating to all facets of energy. Coverage is objective and factual. Close to 1000 sources are reviewed regularly for the most significant and current information on energy topics.

ENERGY SCIENCES
832 Rockville Pike
Rockville, Maryland 20852

Larry Miller
301 278 0988

Solarex retail outlet, recently expanded to include complete water pumping systems, featuring McDonald, March, and Shurflow pumps. Design PV systems and supply all related parts and equipment. An educational catalog, *The Solar Wonderbook*, contains product descriptions and sizing information. Ongoing development and design of new products.

ENERGY SCIENCES
Lake Forest Mall
Gaithersburg, Maryland 20877

Paul Kelly
301 948 3664

ENERGY SCIENCES
Tyson's Corner Center
McLean, Virginia 22102

Greg Klemmer
703 448 8668

THE ENERGY SHOP, INC.
P.O. Box 1512
Victorville, California 92392

D. Buckles, President
619 245 2349

PV systems, tracking devices.

ENERGY SYSTEMS GROUP
Rocky Flats Plant
Box 464
Golden, Colorado 80401

303 497 7000

Microcomputer software for residential/commercial designers in the renewable energy field.

ENERGYWORKS, INC.
44 Hunt Street
Watertown, Massachusetts 02172

617 962 8600

Microcomputer software for residential/commercial designers in the renewable energy field.

ENTECH, INC.
1015 Royal Lane
P.O. Box 612246
DFW Airport, Texas 75261

Robert R. Walters, Vice President – Marketing
214 456 0900

ENTECH manufactures a high-efficiency (>12%) photovoltaic collector that uses a high-efficiency and inexpensive linear Fresnel lens to reduce the amount of silicon cells required by a factor of 40. The collector system uses full two-axis tracking. Collector modules can provide either electricity alone or a combination of electricity and hot water. A demonstration system at the Dallas/Fort Worth Airport, Texas, has continued since June 1982 to perform at the highest efficiency. It delivers up to 27 kW DC electricity and 145 kW of thermal energy. Also, ongoing research and development of low-cost concentrating PV collectors, module and system design improvements.

**ENVIRON ENERGY SYSTEMS
P.O. Box 10998-526
Austin, Texas 78766**

512 250 1072

Produces the model EES-1 photovoltaic closed-loop system for domestic solar water heating.

**ENVIRONMENTAL ALTERNATIVES
818 E. Chestnut Street
Louisville, Kentucky 40205**

Phyllis L. Fitzgerald
502 587 3028

Environmental Alternatives' Urban Alternative Homestead demonstration house offers visitors the opportunity to see alternative energy systems and energy conservation principles at work. In addition to a solar greenhouse, solar cooker and dehydrator, a photovoltaic array and windmill are in operation. An information center, energy library and a variety of workshops are offered.

EPPLEY LABORATORY
12 Sheffield Avenue
Newport, Rhode Island 02840

401 847 1020

Manufacture of measuring instruments.

E-SYSTEMS
P.O. Box 226118
Dallas, Texas 75266

214 272 0515

Manufacturing/R&D—linear Fresnel lens solar concentrators. Photovoltaic thermal concentrators.

EUREKA DESIGN
1955 6th Avenue, West
Seattle, Washington 98119

Ted P. Lehn
206 282 0751

Solar architecture, design, energy management. Supplier of PV products.

EXIDE CORP.
101 Gibralter Road
Horsham, Pennsylvania 19044

215 674 9500

Manufacturer of batteries for PV applications.

EXPORT COUNCIL FOR RENEWABLE ENERGY, U.S.
P.O. Box 1300
Washington, D.C. 20013

202 466 6350

F

F-CHART SOFTWARE
4406 Fox Bluff Road
Middleton, Wisconsin 53562

Sylvia Beckman
608 836 8536

The originators of the F-Chart method, S. A. Klein and W. A. Beckman, have developed an interactive photovoltaic systems analysis program for the APPLE, IBM PC, TRS 80 and most CP/M-based microcomputers. The program, based on methods developed at the University of Wisconsin, can analyze three different systems: battery storage, utility feedback, and stand-alone. The methods rely on solar radiation utilizability to account for statistical variation of radiation and load on system performance. Included are complete economic analysis capability, weather data for 232 U.S. and 97 Canadian cities, graphical output, a HELP command, the ability to store system descriptions for future use, and many other features. A comprehensive manual which describes the algorithms is included with the program or is available separately.

FLAD & ASSOCIATES
6200 Mineral Point Road
Madison, Wisconsin 53705

Richard Hoyord
608 238 2661

Information dissemination to clients involved in major building programs.

UNIVERSITY OF FLORIDA
Engineering & Industrial Experiment Station
300 Weil Hall
Gainesville, Florida 32611

Dr. Wayne H. Chen, Director
904 392 0941

Photovoltaics research.

FLORIDA SOLAR ENERGY CENTER
300 State Road 401
Cape Canaveral, Florida 32920

Linda Carpenter/Ingrid Melody
305 783 0300

The Florida Solar Energy Center is a solar energy research institute of the state university system. The Center tests, rates and certifies solar energy collectors manufactured or for sale in the state.

The Center is the lead organization in the Photovoltaic Southeast Residential Experiment Station, funded by DOE. The SE RES facilities include a utility-interactive residence and five prototype PV residences on the Center's grounds, a Center-sited PV flexible test facility and sun-tracking test stands, and field PV test sites through the southeast. Goal of SE RES is to determine the best systems and designs for utility-interactive PV residences in the southeast.

The Center also performs research on stand-alone PV systems for powering remotely sited traffic control systems, and on PV systems to power electrolyzers that separate water into hydrogen and oxygen.

Educational activities include Practical Photovoltaics Short Course (offered quarterly), Fundamentals & Applications of Photovoltaics (two-day workshops), plus general and technical publications, including a quarterly newsletter.

FLORIDA SOLAR HEATING SYSTEMS, INC.
2976 Matthew Drive
Rockledge, Florida 32955

Steve Murphy
305 631 2255

Supplier of both PV and auxiliary equipment, including products by ARCO Solar, Solarex and Exxon, Decco batteries, and Hartell pumps.

JOHN FLUKE MANUFACTURING CO.
6920 Seaway Boulevard
Everett, Washington 98206

Jerome Froland, Vice President – Marketing
206 356 5310

Manufactures digital multimeters and other electronic metering devices.

KENNETH FOSTER
1742 Dowd Drive
St. Louis, Missouri 63136

314 522 6667

Supplier of solar cells, 1.5, 1.7 and 1.8 amp, plus lower-powered and cracked cells when available, 24- and 65-amp nickel cadmium batteries, and 100-amp hr heavy-duty lead-acid batteries. Product updates and information, with quantity breaks, on request.

FRANKLIN ELECTRIC CO., INC.
402 E. Spring Street
Bluffton, Indiana 46714

219 824 2900

Manufacturer of submersible pump motors.

FREE ENERGY OPTIONS
P.O. Box 430
Veneta, Oregon 97487

Leo D. Morin
503 935 2749

Dealer for ARCO and Solec modules, Heart inverters, and Wagstaff batteries. Also a dealer for a variety of alternative energy appliances, pumps, lights, refrigerators, micro-hydro units, PV charge controllers, stand-by power systems, etc. Price list available on request.

Licensed contractor, installation services available on most projects, along with system and component design.

FREE ENERGY SYSTEMS INC.
Mount & Red Hill Roads
P.O. Box 3030
Lenni, Pennsylvania 19052

Gerry Keenan
215 583 4780

Free Energy Systems Inc. is involved in the manufacture of marine-grade photovoltaic panels. These panels were designed to withstand the harsh conditions encountered onboard pleasure and racing boats, as well as recreational vehicles.

FES Inc. is also active in the custom PV market, designing and manufacturing highly specialized PV panels for the military and civilian communications industries. FES Inc. continues to distribute a complete line of terrestrial PV panels and balance of systems equipment for residential, commercial and agricultural applications.

FROST AND SULLIVAN
106 Fulton Street
New York, New York 10038

212 233 1080

Technology-based research organization. Their recent study predicts 131,000 megawatts of solar electric generating equipment installed worldwide by the year 2010—over 1000 times today's level.

FUJI ELECTRIC CO.
2-2-1, Nagasaka
Yokosuka-shi 240-01
Japan

Manufacturer of amorphous silicon cells for consumer electronic products.

G

GATES ENERGY PRODUCTS
1050 So. Broadway
P.O. Box 5887
Denver, Colorado 80217

Ronald O. Hamel, Vice President – Marketing
C. J. Ebert, Advertising Manager
303 744 4806

Rechargeable, sealed lead-acid batteries which can withstand significant overcharge and voltage regulation variation without outgassing. High efficiency; self-limiting charge characteristics; no need for voltage regulator; wide operating temperature range.

GENERAL ELECTRIC COMPANY
Advanced Energy Programs Department
P.O. Box 13601
Philadelphia, Pennsylvania 19101

215 962 2112

R&D residential PV "shingles." Prototype PV residential design.

GEORGIA POWER CO.
Energy Research Division
333 Piedmont Avenue, N.E.
Atlanta, Georgia 30308

Gary Birdwell, Sr. Specialist
404 526 6526

Public utility active in Utility Research Group (URG), which is investigating the technical aspects of PV/grid interconnections.

DREW A. GILLETT, P.E.
319 Holbrook Road
Bedford, New Hampshire 03102

603 668 7336

Integrated solar and mechanical/electrical consulting services for commercial, residential and institutional projects. Solar feasibility studies; active and passive solar design; daylighting analysis and models; photovoltaic system design; construction supervision; building energy analysis.

GLOBAL PHOTOVOLTAIC SPECIALISTS INC.
22432 DeGrasse Drive
Woodland Hills, California 91364

Howard Somberg, President
818 999 4399

GPS Inc. is a consulting engineering firm that provides services in manufacturing process development, module design, automated equipment design and fabrication, development of OEM applications, and technology transfer of complete solar cell and module production facilities. We also provide services in advanced processes for the manufacture of thin-film devices and monolithic modules. Seventeen years experience in the PV industry. GPS Inc. also manufactures low-power (2 – 10 watts) modules for specialized applications and does systems application analyses.

GLOBE BATTERY
5757 N. Green Bay Avenue
Milwaukee, Wisconsin 53209

414 228 1200

Manufacturer of PV batteries.

GML SYSTEMS, INC.
2175 Kingsley Avenue, No. 316
Orange Park, Florida 32073

Jerry Liverette
904 272 3003

Supplier of PV equipment.

GNB BATTERIES INC.
P.O. Box 43140
St. Paul, Minnesota 55164

Bob Stoll
612 681 5000

PV and wind batteries including the Action Pack deep-cycle battery for stand-alone systems. Stationary cells for uninterruptible power systems.

GOLD STAR ENERGY SAVING PRODUCTS LTD.
102 Packham Avenue
Saskatoon, Saskatchewan S7N 2S6
Canada

G. Hanipel
306 373 6200

We are involved in the sales and distribution of any products relating to the renewable energies and energy conservation.

GPL INDUSTRIES
P.O. Box 306
La Canada, California 91011

Gary Parker
213 956 6603

Manufacturer of the Sunmill line of photovoltaic-powered water pumps, especially suited to low-flow wells and spring boxes but will pump a generous supply for a normal dwelling or for stock watering needs. A photovoltaic panel is coupled by an electronic controller to a DC motor which is part of the Sunmill's jack pump. The jack is specially designed for this application. It includes a directly coupled, enclosed flywheel and a high-efficiency gear box to drive a counterweight balanced crank. No batteries are necessary. Only simple annual maintenance is required.

ARNOLD GREENE TESTING LABORATORIES, INC.
East Natick Industrial Park
Natick, Massachusetts 01760

S. W. Floss
617 653 5950

This laboratory offers testing to industry: (1) physical testing of materials; (2) X-ray, nondestructive testing of all sorts; (3) chemistry; (4) product engineering—hydrostatic testing, acoustic testing, stress analysis; and (5) solid fuel appliances.

ALVIN L. GREGORY
Solar Energy Consultant
5860 Callister Avenue
Sacramento, California 95819

Alvin L. Gregory, Director – Research & Development
916 455 3100

Active in the solar energy field for over 30 years. Research in the photovoltaics segment includes generation of 60-cycle AC electrical power from solar cells without a converter. Also, solar-powered pump using flat-plate collectors, and deep-well pump using solar power.

Research in the generation of electrical power from solar energy through the use of a new type of thermal engine (converting the solar energy to heat energy which activates a new type of heat engine driving an electrical generator). Securing patents on above described engine and pumps.

GRUNDFOS PUMPS CORP.
2555 Clovis Avenue
Clovis, California 93612

209 299 9741

Manufacturer of circulating pumps, submersibles.

GSC ENGINEERED PRODUCTS
4125-R South 68th East Avenue
Tulsa, Oklahoma 74145

W. P. Robey
918 664 1677

Wholesale distributor for Solarex in Oklahoma.

GUMBS ASSOCIATES, INC.
26 Avenue B
Newark, New Jersey 07114

Dr. Ronald W. Gumbs
201 824 5110

Photovoltaics research.

HAENNI INSTRUMENTS INC.
P.O. Box 827
Kenner, Louisiana 70063

504 469 6920

Manufacturer of pressure gauges, thermometers.

BETHE HAGENS/BILL BECKER
105 Wolpers Road
Park Forest, Illinois 60466

312 481 6168

Our PV/wind/solar-powered home is our research site and office/conference center. Richard Komp has done numerous PV educational workshops here. Bill is designing a hybrid sun/wind vehicle prototype to be entered in the cross-country solar rally. Bethe is working on an international trade and communications network with Robert Theobald and Bill Ellis, with quarterly editorials in *Cultural Futures Research* (an anthropology journal).

ROBERT B. HALEY
Box 626
Blacksburg, Virginia 24060

703 951 7811

Microcomputer software for residential/commercial designers in the renewable energy field.

HARBOR FREIGHT SALVAGE
3491 Mission Oaks Boulevard
Camarillo, California 93010

1 800 423 2567

Supplier of 12-volt DC lights. Numerous other items of possible interest to do-it-yourselfers.

HARTELL DIV. MILTON ROY CO.
70 Industrial Drive
Ivyland, Pennsylvania 18974

Douglas Bingler, Applications Engineer
215 322 0730

Manufacturer of PV-powered pumps. The magnetic drive circulator pump series includes electronically commutated, ultra high-efficiency DC motors, engineered to operate from low-wattage photovoltaic panels. The brushless DC circulator pump has been engineered for photovoltaic powered applications; suitable for open or closed loop, water or freeze protected fluid systems.

UNIVERSITY OF HAWAII AT MANOA
Hawaii Natural Energy Institute (HNEI)
2540 Dole Street, Holmes 246
Honolulu, Hawaii 96822

Dr. John W. Shupe, Director
808 948 8890

Research, development and demonstration of alternate energies, including concentrated silicon solar cells.

UNIVERSITY OF HAWAII
Hawaii Natural Energy Institute (HNEI)
Department of Planning & Economic Development
335 Merchant Street, Room 110
Honolulu, Hawaii 96813

D. Richard Neill, PV Program Manager/Renewable Energy Advisor
808 548 4195 or 808 948 8788

Goal is to accelerate the commercial viability of photovoltaic power and its potential contribution to Hawaii's energy future through research, education, workshops, seminars, fact sheets, demonstration projects, *Guidebook on PV Applications in Hawaii*, *PV Update* newsletter, etc.

HAWAIIAN SOLAR ELECTRIC
345 N. Nimitz Highway
Honolulu, Hawaii 96817

Cully Judd
808 523 0711

Distributor in state of Hawaii and the South Pacific area of PV panels and allied products.

HEART INTERFACE
1626 S. 341st Place
Federal Way, Washington 98003

206 838 4295 or 1 800 732 3201

Manufacturer of high-efficiency, 1000-watt, 12-volt power inverters. Optional battery charging capability. Electronic circuit breaker and thermal protection. Anodized aluminum and stainless steel construction. 2.5, 3 and 5 kW models available.

HELIONETICS, DECC DIV.
17312 Eastman Street
Irvine, California 92714

714 546 4731

Manufacturer of inverters, converters. R&D—single-crystal cells.

HELIOTROPE GENERAL, INC.
3733 Kenora Drive
Spring Valley, California 92077

John Blake III
619 460 3930

Heliotrope General has introduced a new line of photovoltaic charge controllers tradenamed HI-ETA for high efficiency. The SMC-1200/2400 is a switch mode power converter and represents the latest advance in two-step battery charge control technology. It is intended for use in stand-alone photovoltaic, wind, or other alternate power generation systems to effectively regulate available energy while providing load management functions. The principal feature of this switch mode controller is its ability to provide a full 30 amps in a regulated as well as a direct charge mode. The control is available in 12 volts DC (model 1200) or 24 volts DC (model 2400).

Also available is the SMC-2 battery voltage regulator. A 12-volts DC solid-state control, it is intended for use in small stand-alone photovoltaic systems. It is designed to efficiently maintain a battery at the correct state of charge and prevent overcharging. Two status LEDs on the front of the control indicate when the control is providing maximum current to the battery load and when the battery is at full charge and being charged at a reduced current level.

HOLEC INC.
11C Esquire Road
North Billerica, Massachusetts 01862

617 667 4120

Packaged PV systems for export. Rural electrification R&D.

HOLLIS OBSERVATORY
One Pine Street
Nashua, New Hampshire 03060

603 882 5017

Manufacturer of pyranometers, integrators, radiometers.

HOME ENERGY WORKSHOP, INC.
1558 Riverside
Ft. Collins, Colorado 80524

Jim Welch
303 482 9507

Home Energy Workshop has been involved with photovoltaics for the last three years. Our expertise is in three areas—education, design and installation. Under the guidance of Dr. Richard Komp, H.E.W. was the first group in the country trained to run his PV workshops. The workshops are oriented to laypeople and include a complete lecture program to introduce how PV works, its design, sizing and applications, and a shop session where participants assemble PV panels from individual cells. We run the PV workshops locally in our shop and for colleges and nonprofit organizations in our region.

Our experience in design and installation is a product of a diverse background among H.E.W. staff members. Our president, Jim Welch, has a Masters Degree from Arizona State University in Solar Design and Technology and has considerable experience in the design and sizing of solar heating and electrical systems. Mike McGoey has run installation crews for PV systems powering remote communications and other remote applications. He is completing his Masters Degree in Industrial Science. Phil Friedman, who conducts our workshops, has an electrical technician's background and the teaching experience to communicate electronic theory to laypeople.

**HOMESTEAD ELECTRIC
P.O. Box 451
Northport, Washington 99157**

Dave Johnson

Manufacturer of control boards, universal panel mounts, and rechargeable flashlights. Supplier of ARCO Solar products, full-spectrum 12-volt DC fluorescent lighting systems, components and batteries. Research ongoing in thermo-electric backup systems. Mobile demonstration unit available.

**HONEYWELL CORPORATION
10701 Lyndale Avenue, South
Bloomington, Minnesota 55420**

612 887 4489

R&D on silicon-on-ceramic cells.

**HOOD MILLER ASSOCIATES
2051 Leavenworth Street
San Francisco, California 94133**

Bobbie Sue Hood, Architect
415 771 7770

Architects and real estate developers that provide full architectural design and construction documents. Research activities include a prototype PV installation in a San Francisco residence.

HORIZON BUILDERS
218 Park Avenue
P.O. Box 1303
Raton, New Mexico 85540

Kenneth D. Sandelin
505 445 5133

Residential designer; PV equipment suppliers (ARCO Solar and others). Teach solar design classes, freshman and sophomore level, at Trinidad State Junior College, Trinidad, Colorado.

HOWARD DESIGN B.V.
Den Biest 22 en 31
5615AV Eindhoven
The Netherlands

Iain F. Garner
31 40 126125

Active in PV systems research—new technologies, investment analysis and applications research. Publication of technical papers at European conferences. Services in PV market research, advertising and press relations including graphic design, photography, audio-visual and video presentations. Origination, translation and adaptation in English, French, Dutch and German. Clients include Honeywell Europe, Bell Telephone S.A., and many divisions of the Philips concern.

HUGHES AIRCRAFT
Solar Energy Department
1550 Hughes Way
Long Beach, California 90810

George J. Naff
213 513 3000

PV installations. Projects include the PV Higher Education Exemplar Facility at Georgetown University, Washington, D.C.

M. HUTTON & CO.
3240 Garden Brook
Dallas, Texas 75234

Mark Wiener/Bryan Martin
214 484 0580 or 1 800 527 0627

M. Hutton Company is an authorized distributor of ARCO Solar photovoltaic modules and support equipment for the southeastern and southwestern U.S. Hutton can size and supply components for solar electric systems and solar hot water systems, both residential and commercial; also specializes in mobile communications equipment and radio service test instruments.

HYDROCAP CORP.
975 N.W. 95th Street
Miami, Florida 33150

George Peroni
305 696 2504

Manufacture gas recombining catalytic battery caps. HYDROCAPS eliminate corrosion, recombine hydrogen gas into harmless water, and greatly reduce watering maintenance. They also provide a handy means for evaluating the battery and charging system. During final charging with a properly operating system, HYDROCAPS are just warm to the touch. If one or all are hot, overcharging is occurring.

HYDROGEN WIND, INC.
R.R. 2, Box 262
Lineville, Iowa 50147

Laurence Spicer
515 876 5665

Manufacturer of hydrogen fuel production equipment which operates on a DC power supply—wind, PV, etc. Products include one-cell electrolyzer (producing 0.75 cubic feet of hydrogen/hour); four-cell electrolyzer (producing 3 cubic feet of hydrogen/hour); hydrogen – oxygen pressure controls (available in three ranges);

and 1-kW, 2-kW and 10-kW electrolyzer systems (producing 10, 20 and 80 cubic feet of hydrogen/hour, respectively). Information packet and prices available on request.

IBE, INC.
9121 De Garmo Avenue
Sun Valley, California 91352

Manufacturer of deep-cycle batteries.

UNIVERSITY OF ILLINOIS
Solid State Electronics Laboratory
Electrical Engineering Research Laboratory
Urbana, Illinois 61803

Dr. Chih-Tang Sah, Director

Semiconductor physics research, PV energy conversion devices.

INDEPENDENT ELECTRIC COMPANY (IECO)
8880 Arapaho
P.O. Box 9309
Casper, Wyoming 82609

Mathew Overeem
307 472 2118 or 307 265 7579

Manufacturer of controls and microprocessor-based controls. Supplier of PV panels and all auxiliary equipment.

We have now begun prototyping a simple water pump package powered by PV and have done extensive research and calculations on all sizes and types of PV pumping, from water to oil.

We are primarily interested in providing power to remote areas, either homesteads or commercial, and have an inhouse staff capable of sizing all remote power needs—rural home to cathodic protection and communications repeater sites. We are concerned with the design and evaluation of hybrid systems of PV and wind electric systems and have developed a simple process to aid in the initial analysis of projects. We also offer seminars and workshops on practical PV.

INDEPENDENT ENERGY PRODUCERS ASSOCIATION (IEP)
1225 8th Street, Suite 285
Sacramento, California 95814

Dr. Jan Hamrin, Executive Director
916 448 9499

INDEPENDENT HOME ENERGY SYSTEMS
1938 Dutch Creek Road
Yreka, California 96097

William Hartzell
916 465 2308

PV dealer serving the Siskiyou County area in northern California. Specializes in remote home solar electric power systems, solar water pumping, solar water heating, and hydroelectric systems.

INDEPENDENT POWER COMPANY
12340 Tyler Foote Road
Nevada City, California 95959

Ron Kenedi/Sam Vanderhoof
916 292 3754

Full-line dealer of PV components, accessories, and low-voltage appliances. Manufacturer of mounts and harnesses, control boxes, 12-volt blenders, ni-cad battery chargers, hydro power plants, and light fixtures. Offer complete system design services for PV and hydroelectric plants. Research activities cover PV systems and applications, water pumping and refrigeration. Also, teach alternative energy courses at Sierra Nevada College. Catalog available on request.

INSERVCO, INC.
114 Commerce Avenue
LaGrange, Ohio 44050

216 355 5102

Solar instrumentation, electronics.

INSOLATION SOLAR
8185 Sunset Drive
Lakewood, Colorado 80215

Benjamin D. Boltz
303 233 3760

Manufacturer of electronic controls.

INSTITUTE OF AMORPHOUS STUDIES
1050 E. Square Lake Road
Bloomfield Hills, Michigan 48013

313 540 0102

INSTITUTE OF ELECTRICAL AND ELECTRONIC ENGINEERS (IEEE)
345 E. 47th Street
New York, New York 10017

212 705 7900

Provides leadership in developing the Standards for PV and for Storage and Generating Systems, which discuss the "power quality" at the connections of the dispersed PV and the grid system.

INTEGRATED POWER
7624 Airpark Road
Gaithersburg, Maryland 20879

301 963 5884

Packaged systems.

INTEGRATED SOLAR SYSTEMS INC.
2915 Commercial Avenue
Anacortes, Washington 98221

Greg Snelson
206 293 9593

Supplier of PV components and complete systems. Design and installation services provided.

INTERSOL POWER CORPORATION
11901 West Cedar Avenue
Lakewood, Colorado 80228

John A. Sanders, President
303 989 8710

Manufacturer of 5.4-kW PV concentrator arrays, PV concentrator modules, mobile PV systems (available in 500, 1000, 1500 and 2000 watts), one megawatt PV Concentrator System and tracking control structures. Services include PV power systems integration and solar/thermal design and installation.

IOTA ENGINEERING CO.
4700 S. Park Avenue, Suite 8
Tucson, Arizona 85714

Harry H. Clayton/Edward Skip Zahm
602 294 3292

IOTA designs and manufactures a line of solid-state ballasts for energy-efficient DC-powered lighting, and a low-voltage high-quality ruggedized fluorescent light (tradename REVL) for recreational and emergency vehicle use. Ballasts (12-volt) and accessories that permit use of a group of low-wattage, very efficient small fluorescent lamps are just being introduced. These products permit the modification of standard household fixtures to operate, and look like, incandescent bulb lights but require only one-sixth the energy. This helps the homeowner to select lighting fixtures that require the least power for illumination.

IOTA has over 15 years experience in the design and manufacture of solid-state ballasts for DC-powered fluorescent lighting, including the development of high-frequency ballasts for special lights used in NASA's SKYLAB. Research activities include solid-state power conditioning products.

IOWA ENERGY POLICY COUNCIL
Energy Hotline
Lucas Building
Des Moines, Iowa 50319

Randy Martin/Jeff Newburn
515 281 7017 or 1 800 532 1114

State energy office and energy information service for Iowa residents.

J

JACUZZI BROTHERS
1151 New Benton Highway
Little Rock, Arkansas 72201

501 455 1234

Manufacturer of pumps for use in PV-powered pumping systems.

JAPAN SOLAR ENERGY CORPORATION
11-17 Kogahonmachi
Fushimiku, Kyoto 612
Japan

PV development: ribbon process production and marketing.

JET PROPULSION LABS
California Institute of Technology
4800 Oak Grove Drive
Pasadena, California 91109

213 354 4321

R&D; testing and evaluating PV modules. Sponsoring lab for numerous PV research projects.

JORDAN ENERGY INSTITUTE
155 Seven Mile road
Comstock Park, Michigan 49321

Conrad Heins
616 784 7595

AE470, Design 2: An upper-level college course that includes photovoltaic theory, manufacturing processes, recent research results, and the sizing and design of PV power systems. Also offers PV workshops and maintains an energy library.

K

KALAMAZOO ENERGY OFFICE
7000 N. Westnedge Avenue
Kalamazoo, Michigan 49007

Mark H. Clevey
616 381 1574

PV advocacy in Michigan. Mr. Clevey is an author and photovoltaics course instructor, Lansing Community College, Lansing, Michigan.

KANSAS UNIVERSITY SOLAR ENERGY CLUB
P.O. Box 13 Union
Lawrence, Kansas 66045

James Mendenhall
913 864 4589

Founded in 1979, we have established a 1500-volume energy library through donations. We also offer hands-on workshops. Currently planning design award competition for PV-powered standing clock for a special on-campus location.

KAYEX CORPORATION
1000 Millstead Way
Rochester, New York 14624

Dr. Richard L. Lane, Director of Technology
716 235 2524

Manufacturer of crystal growth and wafer slicing equipment for silicon and gallium arsenide cells. This includes HAMCO large-capacity (up to 100 kg) crystal growing furnaces, Capco slicing equipment for all wafer sizes up to 6 inches, and Spitfire lap and polishing system.

KECK & KECK, ARCHITECTS
612 N. Michigan Avenue, Room 502
Chicago, Illinois 60611

William Keck, F.A.I.A.
312 787 5035

Architectural firm which researches and utilizes good design applications of solar energy systems.

KEE INDUSTRIAL PRODUCTS, INC.
P.O. Box 207
Buffalo, New York 14225

Michael G. Millard
716 685 8250

Manufactures slip-on pipe fittings and bolt adapter systems for construction of structural framework of solar collector support systems. Free 20-page catalog available.

KENNING
314 S. Morton Street
Waupaca, Wisconsin 54981

Kenneth L. Verhalen

715 258 5148

Design and construction (passive and active, including PV) of residential structures. Personnel management services.

KINGSTON INDUSTRIES CORP.
Solar Products Division
199 S. Main
Liberty, New York 12754

914 292 6020

Manufacturer of King-Lux solar reflector sheet, high-purity aluminum, for reflector booster for PV cells and other solar collectors.

KOMATSU LTD.
3-6, 2-Chome
Akasaka, Tokyo
Japan

584 7111

Advanced R&D in amorphous silicon solar cells—recently reported 10.7% efficiency.

KULICKE & SOFFA IND.
104 Wittmer Road
Horsham, Pennsylvania 19044

215 674 2800

Module assembly machines.

KYOCERA INTERNATIONAL
8611 Balboa Avenue
San Diego, California 92123

619 576 2648

North American photovoltaic product development and marketing for Japan Solar Energy Corporation.

L

WM. LAMB CO.
10615 Chandler Boulevard
North Hollywood, California 91601

Joel Davidson
213 980 6248

Distributor for ARCO Solar products. Manufacturer of PV-powered deep-well pumping systems, PV-powered evaporative cooler (four models available, including an RV unit and a window unit), and PV-powered ceiling fan packages.

PETER T. LANDSBERG
Visiting Professor, Electrical Engineering
University of Florida
Gainesville, Florida 32611

Research—theory of photovoltaics and energy conversion. To continue at the University of Southampton, S09 5NH, England (559122/556545).

LANE ENERGY CENTER
2915 Row River Road
Cottage Grove, Oregon 97424

Bob Mieger
503 942 0522

PV supplier—components and systems. Technical assistance and economic analysis services.

LAST CHANCE HOMESTEADS
P.O. Box 31480
El Paso, Texas 77931

Stephen Michael Hensley
915 757 2611 or 915 757 0189

Design, install and maintain remote location PV and DC home systems. Specialize only in off-grid systems.

LENBROOK INDUSTRIES LIMITED
1145 Bellamy Road
Scarborough, Ontario M1H 1H5
Canada

Donald E. Simmonds, Manager – Energy Division
416 438 4610

Lenbrook Industries is a Canadian owned company specializing in photovoltaic systems and telecommunications equipment. Supplier and installer of Solarex photovoltaic systems in Canada, Lenbrook is now assembling Solarex solar modules in Canada using Semix polycrystalline solar cells.

LEVELEG PRECISION SOLAR MOUNTING SYSTEMS
8656 Commerce
San Diego, California 92121

V. Dennis, President
619 271 6240

Roof mounting systems for thermal collectors and PV systems.

LICOR INC.
P.O. Box 4425
Lincoln, Nebraska 68504

Manufacturer of pyranometers and integrators.

ARTHUR D. LITTLE
P.O. Box 15-117A
Cambridge, Massachusetts 02140

617 864 5770

R&D—ribbon and sheet growth cells; testing.

LONDE, PARKER, MICHELS
150 N. Meramec, Suite 205
St. Louis, Missouri 63105

314 725 5501

Microcomputer software for residential/commercial designers in the renewable energy field.

LONG ISLAND SOLAR ENERGY ASSOCIATION
7 Sentinel Place
Massapequa, New York 11758

Al Lewandowski/Steven Kvit
516 694 4752 or 516 798 5351

Activities include consumer education through hands-on construction workshops; research to increase efficiency; energy expositions; monthly educational seminars and grass roots meetings; speaking engagements utilizing films and slide shows.

LOS ALAMOS NATIONAL LABORATORY
P.O. Box 1663
Los Alamos, New Mexico 87545

William Keller
505 667 5061

R&D—magnetic refrigeration.

LUXTRON INC.
241 Winter Street
Haverhill, Massachusetts 01830

William Slusher, Engineer
617 372 5211

Manufacturer of UPS backup systems.

MAINE OFFICE OF ENERGY RESOURCES
State House No. 53
Augusta, Maine 04333

Brian Kent/Richard Bay
207 289 3811

Informational activities include: production of a booklet entitled "Electricity from the Sun"; 30-minute video documentary on PV use on Maine's Monhegan Island; PV education workshops; products/appliances/dealer files; PV slide program library.

MARCH MFG. INC.
1819 Pickwick Avenue
Glenview, Illinois 60025

Fred Ahline/Fred Zimmermann
312 729 5300

Manufacturer of 12-volt DC solar/hydraulic pumps, developed specifically for use with PV.

MARTIN MARIETTA DENVER AEROSPACE
Solar Energy Systems
P.O. Box 179, M.S. L0450
Denver, Colorado 80201

R. L. Parker
303 977 0107

Manufacturer of concentrating PV systems and supplier of PV turnkey systems. Designed Phoenix's Sky Harbor 225-kWp concentrating PV plant, grid-connected to Arizona Public Services (APS) and activated April 1982. Through SOLERAS, a Saudi Arabia–U.S. solar energy cooperative program, designed and built a 350-kWp system near Riyadh, electrifying three villages. Also built the International Energy Agency's 500-kW system, Almeria, Spain.

**MARVEL DIVISION
DAYTON-WALTHER CORP.**
P.O. Box 997
Richmond, Indiana 47374

317 962 2521 or 1 800 428 6644

Manufacturer of DC refrigeration equipment and conversion kits.

MASSACHUSETTS INSTITUTE OF TECHNOLOGY
Designers Software Exchange
Laboratory of Architecture and Planning
77 Massachusetts Avenue
Cambridge, Massachusetts 02139

617 253 1000

Library of energy-related software. Accepted programs meet stringent criteria. Catalog and updates available.

MASSACHUSETTS INSTITUTE OF TECHNOLOGY
MIT Lincoln Laboratory
244 Wood Street
Lexington, Massachusetts 02173

Dr. F. J. Solman
617 863 5500

Research large-scale PV systems, including the 100-kWp Natural Bridges National Monument test system.

MASSACHUSETTS INSTITUTE OF TECHNOLOGY
Energy Laboratory, Room E40-455
Cambridge, Massachusetts 02139

Prof. David C. White, Director
617 253 3400

Examines social and technological issues involved in energy consumption; assesses alternative energies emphasizing economics and politics of international trade.

MASSDESIGN ARCHITECTS & PLANNERS, INC.
146 Mt. Auburn
Cambridge, Massachusetts 02138

617 491 0961

Architectural design incorporating PV.

K. L. MATHERS & ASSOCIATES
Route 3, Box 217
Momence, Illinois 60954

Kris L. Mathers, President
815 944 5385

K. L. Mathers & Associates serves residential, commercial and industrial clients in the areas of energy conservation and solar energy system consultation. Energy conservation services include: energy appraisals with both technical and financial evaluations and recommendations; transportation energy analysis and educational training programs in energy conservation. Mr. Mathers has been teaching college-level courses and seminars in alternative energy systems design, photovoltaics, and energy conservation since the mid-1970s.

MATSUSHITA ELECTRIC INDUSTRIAL CO., LTD.
(Panasonic)
Kadoma, Osaka 571
Japan

Development of screen-printed thin-film cells.

A.Y. McDONALD MFG. CO.
4800 Chavenelle Road
Dubuque, Iowa 52001

John Eckel
319 583 7311

Manufacturer of PV panel – direct water pumping systems and PV-rechargeable battery-powered pumping systems. Both systems offer dependable, cost-effective delivery of drinking and irrigation water where conventional power sources are unavailable. The first system provides DC current direct from PV panels to the pump, for dependable operation during sunlight hours. The second system uses PV rechargeable deep-cycle storage batteries to provide DC power on demand any time, day or night.

MC SOLAR ENGINEERING
19675 Skyline Boulevard
Los Gatos, California 95030

Michael Clifton
408 395 4848

Design and installation of photovoltaic systems.

RICK McGOWAN
362 Main Street
Burlington, Vermont 05401

802 658 3890

Installation of PV-powered instrumentation for passive solar home monitoring, windmill monitoring, village water supply, and cold chain vaccine refrigeration units. Have worked extensively overseas. Design projects include PV-powered data loggers, anemometry, pumping and refrigeration systems.

MECHANICAL PRODUCTS INC.
P.O. Box 729
Jackson, Michigan 49204

517 782 0391

Manufacturer of 12-volt DC circuit breakers.

LENA MENASHIAN
Information Consultant
P.O. Box 454
Mountain View, California 94042

415 961 8589

Provides information services and training support for the photovoltaics industry. Services include: research support; design and teaching of PV workshops/seminars (consumer-oriented, adult level); and information support through analysis of needs, automated or manual retrieval, organization and dissemination of required information.

MERIDIAN CORPORATION
5201 Leesburg Pike, Suite 400
Falls Church, Virginia 22041

Pete Borgo
703 998 0922

PV and wind applications assessment services. Consulting, resource assessments, siting studies, system sizing, cost–benefit analyses. Specialize in providing independent computerized analysis for small and intermediate size renewable energy applications.

TIM MERRIGAN, P.E.
8702 Jasmine Court
Cape Canaveral, Florida 32920

305 783 0300

Consulting engineer, designs solar thermal systems—both DHW and space heating; stand-alone and grid-connected photovoltaic systems. Building design—envelope and mechanical. Mr. Merrigan is also a research engineer in the R&D division of Florida Solar Energy Center.

METAL MASTERS INC.
14410 E. Nelson Avenue
City of Industry, California 91744

Manufacturer of 12-volt DC PV sunlight converters.

METHODE ELECTRONICS
5633 W. 99th
Chicago Ridge, Illinois 60415

Manufacturer of PV and wind batteries and cables.

MICHIGAN ENERGY ADMINISTRATION
Energy Clearinghouse
100 S. Pine, North Tower, 3rd Floor
P.O. Box 30228
Lansing, Michigan 48909

Mary Jo Gliniecki
517 373 0480 or 1 800 292 4704

Offers numerous publications on all aspects of the renewable energies and energy conservation. Energy Hotline for Michigan residents.

MICROCOMPUTER DESIGN TOOLS
2513 Kimberly Court, N.W.
Albuquerque, New Mexico 87120

505 831 3911

Microcomputer software for residential/commercial designers in the renewable energy field.

MINNESOTA MINING & MANUFACTURING (3M)
3M Center
St. Paul, Minnesota 55144

612 733 1110

R&D to advance state-of-the-art of single-junction, thin-film amorphous silicon solar cells fabricated by glow discharge deposition.

MISSISSIPPI COUNTY COMMUNITY COLLEGE
P.O. Box 1109
Blythesville, Arkansas 72315

C. M. Benson
501 762 1020

Two-year community college offers AAS Degree in Solar Technology. Facilities include large DOE photovoltaic/thermal power project as well as other examples of solar applications and energy conscious design. Photovoltaics power the energy system, producing 5500 kW-hr of energy per day, with the excess being transferred to the utility grid.

MISSOURI DEPARTMENT OF NATURAL RESOURCES
Division of Energy
P.O. Box 176
Jefferson City, Missouri 65101

John Kirby/Howard Hufford
314 751 4000

We provide general PV information to Missouri residents. We also set up seminars and workshops on all solar-related subjects, including PV. We promote solar technologies throughout the state.

MITSUBISHI ELECTRIC COMPANY
4-1, Mizuhara, Itami-Shi
Hyogo 664
Japan

R&D of gallium arsenide cells, Fresnel lenses, and tracking systems.

MOBIL SOLAR ENERGY CORPORATION
16 Hickory Drive
Waltham, Massachusetts 02254

Ann Ellis
617 890 1180

Manufacturer of photovoltaic modules and all balance of systems components, Mobil Solar Energy Corporation (MSEC) is engaged in the commercialization of ribbon photovoltaic modules and complete systems. The MSEC "Ra" family of modules incorporate advanced silicon ribbon cells manufactured by the edge-defined film-fed growth (EFG) process. MSEC will supply complete custom-designed systems for a variety of applications, including remote communications, cathodic protection of wells and pipelines, agricultural applications such as irrigation and water processing (including desalination), and power for remote lighting, small communities, isolated or grid-connected residences.

MONEGON, LTD.
4 Professional Drive, Suite 130
Gaithersburg, Maryland 20879

Peter Grambs
301 258 7540

Monegan, Ltd. specializes in renewable energy systems and energy management products and services. Our services include design and engineering, installation and checkout, technological assessment, energy market analysis, energy audits, energy management and analysis, and publications. Our product line includes photovoltaics, solar thermal panels, solar thermal systems for residential and commercial applications, PV control devices, energy management systems, and publications on renewable energy technologies.

MONEGON, LTD.
15932A Shady Grove Road
Gaithersburg, Maryland 20877

William Barse
301 258 9851

MR. SUN, INC.
1811 Aga Drive
Alexandria, Minnesota 56308

Gary Herrlinger
612 763 3606

Mr. Sun supplies photovoltaics to schools and school supply catalogs for use in classroom environments. We provide individual kits or kits packaged for multiple (10-student) projects. Mr. Sun has published a workbook on PV for classroom use. Our primary level is grades 6 – 12.

MUELLER ASSOCIATES, INC.
1401 S. Edgewood Street
Baltimore, Maryland 21227

Tom Timbario/James S. Moore
301 646 4500 or 301 621 1632

Consulting and engineering of commercial, industrial, and multi-family PV-powered residential structures, including power systems design, electrical layout of arrays, utility grid/load interface, system integration into building, support structure design, instrumentation, specifications development, and code and safety issues.

PV projects include the Copley Energy Denton Photovoltaics Station, Denton, Maryland; Solarex Photovoltaics Breeder, Frederick, Maryland; PV power systems for the MIT and NMSEI prototype residences; George M. Hartley Trustee House, Connecticut; House Office Building Annex 2, Washington, D.C., solar energy system, incorporating PV; Nonfossil Liquid Hydrogen Production Study—E:F Technology.

N

NASA LEWIS RESEARCH CENTER
21000 Brookpark Road
Cleveland, Ohio 44135

William Bifano, Manager – Solar Program
216 433 4000

Research and field testing of stand-alone PV systems.

NATIONAL ASSOCIATION OF HOME BUILDERS
15th and M Streets, N.W.
Washington, D.C. 20005

202 822 0254

NATIONAL ASSOCIATION OF PLUMBING, HEATING AND COOLING CONTRACTORS
1016 20th Street, N.W.
Washington, D.C. 20036

202 331 7675

NATIONAL ASSOCIATION OF SOLAR CONTRACTORS
236 Massachusetts Avenue, N.W., Suite 610
Washington, D.C. 20002

202 543 8869

NATIONAL CLIMATIC CENTER
Federal Building
Asheville, North Carolina 28801

704 258 2850

NATIONAL TECHNICAL INFORMATION SERVICE
U.S. Department of Commerce
5285 Port Royal Road
Springfield, Virginia 22161

703 557 4650

NATURAL POWER, INC.
Francestown Turnpike
New Boston, New Hampshire 03070

J. Enrique Toro
603 487 5512

Natural Power manufactures four instruments that are particularly well-suited to PV systems: (1) Integrating Pyranometer, Model S62 — records solar insolation; (2) Dynamic Loading Switch, Model C20 — voltage-sensitive power relay, can be used in conjunction with or in replacement of a voltage regulator; (3) Amp-Hour & Watt-Hour Meters, Models A35 and A40, respectively — records total power being generated or stored by the system (also offer instantaneous readout of amps and watts); (4) Strip Chart Recorders, Model R10/20 — visual, time-based display of operating parameters. Our company can also supply complete PV systems in the areas of refrigeration and rural homes.

NATURAL SYSTEMS, INC.
14465 Lakeshore Drive
P.O. Box 1956
Clearlake, California 95422

Phil Wilcox
707 994 1311

We are a full-line alternative energy retail store, selling products dealing mainly with solar, wood, wind and water. Our PV systems include remote home power, water pumping, and recreational vehicle battery charging (RVs, boats and airplanes).

NEW ENGLAND PHOTOELECTRIC POWER COMPANY (NEPEPCO)
R.D. 4, Box 295
West Brattleboro, Vermont 05301

Carol Levin
802 254 4670

Distributor for PV panels and systems, BOS. Affiliated with Sunnyside Solar.

NEW ENGLAND SOLAR ENERGY ASSOCIATION
14 Green Street
Box 541
Brattleboro, Vermont 05301

Alex Wilson, Executive Director
802 254 2386

NESEA is a nonprofit membership organization founded in 1974 for the purpose of furthering the understanding and use of solar energy and other renewable energy sources. The Association publishes *Northeast Sun*, a bi-monthly magazine of regional solar news, activities, and technologies; sponsors major technical conferences; serves the needs of professionals and homeowners through educational seminars and workshops; supports the solar industry through lobbying and public education; and serves as a solar clearinghouse for the New England area. NESEA is a regional chapter of the American Solar Energy Society and has 16 local chapters with a combined membership of 3000.

NEW JERSEY DEPARTMENT OF ENERGY
Alternative Technology Unit
101 Commerce Street
Newark, New Jersey 07102

Richard Brandt
201 648 4914

Information dissemination—status of proposed and existing installations; utility plans for involvement; New Jersey corporations (Chronar, RCA, etc.).

NEW MEXICO SOLAR ENERGY INSTITUTE
Box 3-SOL
New Mexico State University
Las Cruces, New Mexico 88003

John F. Schaefer
505 646 4240

Primary research activity is system and component testing and evaluation. Capability in device technology. Design stand-alone and utility-interactive systems, 10 watts to 100 kilowatts. Operate the Southwest Residential Experiment Station for the U.S. Department of Energy. Expertise in hybrid (wind/PV) energy systems, PV-fuel cell interaction, remote data acquisition, PV water pumping. Develop diagnostic systems. Conduct PV workshops nationally; capable of more intensive training. Information dissemination—brochures, fact sheets, manuals—through Solar Information Services, phone 505 646 2651.

NEW VOLT SOLAR ELECTRIC
P.O. Box 36689
Tucson, Arizona 85740

Robert Alan Hershey
602 622 3244

Design and engineering; supplier of PV components. We are currently designing solar homes for photovoltaic applications using the foremost technology available. We also aid others to bid their own projects. Research activities: 12- and 24-volt applications.

NEW YORK STATE ENERGY OFFICE
Agency Building 2
Empire State Place
Albany, New York 12223

Vicki Mastaitis/Linda Brunt
518 473 0729 or 518 474 0521

Information dissemination.

NIPPON ELECTRIC COMPANY, LTD.
1753 Shimonumabe
Nakahara-ku
Kanagawa 211
Japan

Producer of single-crystal silicon cells and modules for remote power systems.

NORCOLD INC.
1503 Michigan Street
Sidney, Ohio 45365

513 492 1111

Manufacturer of DC and RV refrigerators.

NORDIKA SYSTEMS, INC.
7745 E. Redfield
Scottsdale, Arizona 85260

J. J. Garcia, President
602 948 8003

Manufacturer of solar cooling systems.

NORTHCOAST SOLARWORKS
P.O. Box 4209
Arcata, California 95521

Mike Manetas
707 839 3779

Basic consulting, design and installation of PV systems, as well as passive space and water heating. Mr. Manetas also teaches alternative energy courses at Humboldt State University.

NORTHEAST SOLAR ELECTRIC CO.
P.O. Box 184
Endicott, New York 13760

Tom Petruzzelli
607 754 7767

We design photovoltaic systems for industry, communications, remote homeowners, and water pumping. Our markets include the above plus marine, exporters/importers, energy dealers. We are distributors for Solarex and Photowatt, and we supply a large number of support equipment, such as inverters, batteries and appliances. Catalog available.

NORTHERN CALIFORNIA SOLAR ENERGY ASSOCIATION
P.O. Box 886
Berkeley, California 94701

Debra Carroll, Editor/Dale Fousel, President
415 843 4306

NCSEA is a nonprofit organization whose purpose is to foster the development and application of solar energy through the exchange of information. This exchange is accomplished primarily through bi-monthly meetings, open to the public, and through the bi-monthly publication: *The Northern California Sun*. NCSEA is a chapter of the American Solar Energy Society (ASES) and is supported by NCSEA members and advertisers.

NORTH WIND POWER CO.
P.O. Box 556
Moretown, Vermont 05660

J. H. Norton, Jr., President
802 496 2955

Engaged in the design, development, manufacturing and marketing of energy systems, hybrid system controls, wind energy conversion systems. Current projects include a PV/wind system at Bodie Island, North Carolina.

NOVA ELECTRIC MFG. CO.
263 Hillside Avenue
Nutley, New Jersey 07110

201 661 3434

Design, engineering and manufacturing of power conversion equipment: uninterruptible power systems (UPS), inverters and frequency changers.

NOW DEVICES
7975 E. Harvard Avenue, Unit E
Denver, Colorado 80231

303 758 2828

Manufacturer of PV regulators.

OKLAHOMA STATE UNIVERSITY
Institute for Energy Analysis
502 Engineering, North
Stillwater, Oklahoma 74078

Dr. Joe H. Mize, Director
405 624 5700

PV research. Analysis of energy systems, regulations and policies.

OMEGA ELECTRONICA S.A.
J. B. Ambrosetti 491
1405 Buenos Aires
Argentina

Ricardo A. Petrella, President
981 1080/1180/1280

Our firm is the exclusive distributor in the Argentine Republic of ARCO Solar, Inc., as well as of Best Energy Systems. Since 1979 we have been engaged in the design of photovoltaic systems for applications with the military forces, state-owned companies, private companies and domestic applications. We supply complete systems, including panels, regulators, support structures, battery banks, inverters, different types of pumps, fluorescent lamps and accessories in general.

OREGON STATE UNIVERSITY
Energy Extension Service
950 W. 13th Avenue
Eugene, Oregon 97402

Bruce Sullivan, Energy Agent
503 687 4243

Course offering: Photovoltaics: Solar Electricity — overview of residential photovoltaics with a focus on remote applications. Topics include cell technology, system components, efficient energy use, system sizing and cost. Also engage in public information dissemination.

ORIEL CORP.
15 Market Street
P.O. Box 1395
Stamford, Connecticut 06902

203 357 1600

Solar simulators.

P

PARADISE POWER COMPANY
8635 Encino Avenue
Northridge, California 91325

Dr. Lee Morsell
213 349 3701

Manufacturer of the Energy Maximizer, a solid-state peak-power tracker for PV systems. Dr. Lee Morsell, who designed the Energy Maximizer, is currently designing a high-efficiency, submersible, reciprocating positive displacement pump.

PEACHTREE ASSOCIATES, INC.
Box 1312
Decatur, Georgia 30031

404 373 3000

Microcomputer software for residential/commercial designers in the renewable energy field.

UNIVERSITY OF PENNSYLVANIA
Energy Center
Philadelphia, Pennsylvania 19104

Dr. Stephen Feldman, Director
215 898 7185

Multidisciplinary studies, covering technological, social and management facets of photovoltaics and other alternative energies.

BILL PERLEBERG
24110 U.S. Highway 40
Golden, Colorado 80401

303 526 0142

12-volt DC water heater elements.

PHOTOCOMM INCORPORATED
7745 East Redfield Road
Scottsdale, Arizona 85260

Richard C. Cummins
602 948 8003

Manufacturer of photovoltaic modules, balance of systems, and water pumps. Services include systems engineering and design for AC and DC water delivery and stand-alone electric and line-tie; systems sizing; dealer support; and photovoltaic marketing/consulting. With BOSS, recently acquired the manufacturing and supply capabilities of Photowatt International.

PHOTOELECTRIC, INC.
7038 Convoy Court
San Diego, California 92111

Jack K. White, President
619 292 7811

Manufacturer of SI 3000 Solar Inverter and the TDU Controller, available alone or included in the pre-assembled module (Control Central).

PHOTON ENERGY
13 Founders Boulevard
El Paso, Texas 79906

915 779 7774

Manufacturer of PV cells and arrays.

PHOTOVOLTAIC POWER SYSTEMS
8031 Riata Drive
Redding, California 96002

Larry Lourenco
916 365 7263

Design, engineer and install PV power systems as sole source or supplemental energy.

PHOTOVOLTAICS—THE SOLAR ELECTRIC MAGAZINE
P.O. Box 3269
Scottsdale, Arizona 85257

Robert Arganbright
602 829 8167

Technically oriented, international scope periodical.

PHOTOWATT INTERNATIONAL, INC.
2414 W. 14th Street
Tempe, Arizona 85281

Grey Lane
602 894 9564

R&D of amorphous cell program. Recently sold manufacturing and supply sectors to Photocomm and BOSS.

PHOTRON, INC.
149 North Main Street
Willits, California 95490

Vicki Wallace
707 459 3211

Manufactures and distributes alternative electrical source power systems and related equipment. Supplies complete systems and components, *e.g.*, deep-cycle batteries and chargers; manufactures balance of systems; custom designs systems to customer specifications. Maintains a fully instrumented 1-kW PV technical research and design facility for industry testing and development. Holds mobile educational seminars for the public (beginner level), and for its dealers and customers (intermediate and advanced).

PNG CONSERVATION
P.O. Box 37130
Charlotte, North Carolina 28237

Supplier of PV components and auxiliary equipment, offering design and information services.

POCO POWER CORPORATION
4444 Orcutt Road
San Luis Obispo, California 93401

Robert L. Stern
805 543 4444

Manufacturer of balance of systems, also involved in research and design activities. First small power producer (PV) selling power to Pacific Gas & Electric Co., startup August 1982.

POLAROID CORPORATION
549 Technology Square
Cambridge, Massachusetts 02139

617 864 6000

R&D – advanced multijunction amorphous silicon alloy solar cells.

POLAR PRODUCTS
2909 Oregon Court, C-1
Torrance, California 90503

Arthur D. Sams
213 320 3514

Solar-powered refrigerators, lights and water pumps for rural living. Polar Products is offering a line of photovoltaic-powered refrigeration systems not only providing refrigeration and freezing capabilities, these units can supply power for lights, stereos, fans, etc. The water pumps are ideal for household use or for farming of high cash crops. Very practical for drip irrigation systems.

Our current work is in third world village development: refrigeration for vaccine storage, educational services, water pumping, food storage and distribution.

Solar-Powered Refrigerator/Freezer Model RR-2: NASA and World Health Organization tested and recommended; operates from a small PV array and other souces of DC or AC power such as wind and hydro; incorporates a power load diversion system; when both the battery charging and refrigeration power demands are satisfied, the input power is automatically switched to an optional secondary bank of batteries. Export quantity pricing.

POLYDYNE
1230 Sharon Park Drive, Suite 61
Menlo Park, California 94025

415 854 7844

R&D thin-film polycrystalline metallurgical-grade cells.

POPO AGIE, INC.
P.O. Box 71
McKinley Park, Alaska 99755

907 683 2264

Full-line distributor of ARCO Solar products.

POWER PAK
6232 N. Pulaski Road
Chicago, Illinois 60646

David Hibbein
312 286 5566

Full-line distributor for Gates Energy Products.

RICHARD PRATT
P.O. Box 506
Columbus, North Carolina 28722

704 859 6884

Supplier and installer of solar systems; ARCO Solar dealer.

PRIME ENERGY PRODUCTS, INC.
240 N. Pryor Avenue
Roseville, Minnesota 55113

Al Throckmorton, Director of Engineering
612 636 3797

Supplier of Solarex Ventures Group products: panels, fans, batteries, fence chargers, educational and experimental kits, and novelties.

PRINCESS AUTO
P.O. Box 1005
Winnipeg, Manitoba R3C 2W7
Canada

1 800 665 8685

Catalog of tools, electrical and other equipment for the homesteader.

PRINCETON ENERGY GROUP
575 Ewing Street
Princeton, New Jersey 08540

609 921 1965

Microcomputer software for residential/commercial designers in the renewable energy field.

PUBLIC SERVICE COMPANY OF NEW MEXICO (PNM)
Alvarado Square
Albuquerque, New Mexico 87158

R. Frank Burcham, Jr.
505 848 2729

Research on PV/utility intertie, power production/quality monitoring. PNM research projects include: (1) Eldorado Photovoltaic Home — 3-kW roof-mounted flat plat array, near Santa Fe, New Mexico. Utility interfaced. (2) Kirtland Air Force Base (KAFB) Photovoltaic Retrofit Home — 2-kW roof-mounted flat plate array at Kirtland AFB residential area in Albuquerque, New Mexico. Home is intertied with KAFB electrical grid, which is in turn fed by PNM.

PULSTAR CORPORATION
619 South Main Street
Gainesville, Florida 32601

Tom Lane
904 373 5707

OEM manufacturer of photovoltaic modules for domestic solar hot water systems and small commercial systems that can be used with up to 128 square feet of collector area and 240 gallons of storge capacity. Open and closed loop modules. R&D, DHW pumping systems.

PV ENERGY SYSTEMS
2401 Child's Lane
Alexandria, Virginia 22308

Paul Maycock
703 780 9236

Publishes *PV News*, a monthly newsletter covering the technology and its applications, PV investments, utility activity, legislation, homeowner experience, international developments. Mr. Maycock is also an author and consultant.

PVI INC.
P.O. Box 370
Boston, Massachusetts 02117

Tom Warner
617 267 2325

Manufacturer of PV test equipment. Photovoltiac Curve Tracers and auxiliary data acquisition equipment. Custom and standard designs available.

PHOTOVOLTAICS EDITION

PV INSIDERS REPORT
1011 West Colorado Boulevard
Dallas, Texas 75208

Richard Curry, Editor
214 943 4123

Monthly newsletter examines entire photovoltaic field, nation- and worldwide. Covers new products, applications, government activities, industry news, trends, research, PV stocks, etc.

PVI PUBLISHING, INC.
PHOTOVOLTAICS INTERNATIONAL Magazine
2250 North 16th Street, Suite 103
Phoenix, Arizona 85006

Mark C. Fitzgerald, Editor
602 253 9119

PVI Publishing, Inc., publishes the bi-monthly *PHOTOVOLTAICS INTERNATIONAL Magazine* to provide a central source of information for the photovoltaics industry worldwide, on the applications, personnel and programs of the industry. In addition, PVI Publishing, Inc., sponsors and organizes seminars for the photovoltaic industry on marketing, advertising and information dissemination.

PV POWER CORP.
P.O. Box 880
Sebastopol, California 95472

Eugene Brown/Gary Starr
707 525 1111

Design, construction, installation of independent power systems. Large-scale, utility-grade, utility-interface, turnkey installations. Solarex dealer.

Q

QUANTUM ENERGY SYSTEM TECHNOLOGIES
QU.E.S.T.
P.O. Box 9649
Washington, D.C. 20016

Robert Gilbert
202 244 3858

Quantum Energy System Technologies is researching a patent-pending amorphous-based process technology for PV and full semiconductor applications. Ecotech, Inc., the sponsoring organization, provides administrative and marketing research services for QU.E.S.T.

R

RAYS ENERGY CONSULTANTS
1751 N. Grand Avenue, West, No. 41
Springfield, Illinois 62702

217 544 2434

Solar design, calculations, audits, insulation design, educational programs. Two computer systems on-line, including a portable system for onsite demonstration.

RCA CORPORATION
David Sarnoff Research Center
P.O. Box 432
Princeton, New Jersey 08540

R&D on amorphous silicon cells.

READING & BATES DEVELOPMENT COMPANY
3800 First National Building
Tulsa, Oklahoma 74103

918 583 8521

Recently joined Boeing in forming Solvolco to develop thin-film solar cell modules for commercial markets.

REAL GOODS TRADING COMPANY
Alternative Energy Division
308 East Perkins Street
Ukiah, California 95482

Jim Cullen, General Manager
707 468 9212

Real Goods Trading Company has three retail outlets in northern California, as well as a worldwide mail order business. We specialize in the design and installation of PV systems for all applications, and have developed a unique photogen hybrid system. We are presently in the process of completing the latest edition of our Alternative Energy Catalog, which will be a sourcebook of products and system design for the consumer.

REAL GOODS TRADING COMPANY
358 South Main Street
Willits, California 95490

REAL GOODS TRADING COMPANY
5338 West Highway 12
Santa Rosa, California 95401

**REC SPECIALTIES, INC.
530 Constitution Avenue
Camarillo, California 93010**

Patricia Cramer/David Syzmanek
805 987 5021

We manufacture U.L. listed, low amp draw, 12-volt DC fluorescent lights. Over 20 different models are available. They range from an 8-watt model (0.7 amps) to a 40-watt model (2.8 amps). Weatherproof models are available for outdoor use. The electronic inverter ballasts used in these fluorescents were designed by, and are manufactured by, REC Specialties, Inc., and they are available separately. Other voltages available on request.

We also manufacture 12- and 24-volt DC low-pressure sodium lights, 18- and 35-watt. These LPS lights have cast aluminum housings, a vandal-resistant polycarbonate diffuser, and are made standard with a photocell and a timer that is adjustable from 1 to 15 hours.

We manufacture electronic, germicidal water purification systems, which we market under the name of Clean Water Systems of America. These systems have a high-intensity ultraviolet lamp that effects a 99.9+% bacteria, virus and amoeba kill ratio. The consumer model comes fully equipped with a charcoal filter for taste and odor control, mounting hardware, and a dispensing faucet. It draws 2 amps when in operation (at 12 volts DC) and delivers 30 gallons of fully treated water per hour. Commercial and industrial models are also available for low-voltage operation, with the largest system treating 9 gallons per minute. We have the capability to build to customer specifications.

**RED WIND ENERGY EDUCATION CENTER
Highway 61 West
Red Wing, Minnesota 55066**

Pat Enz
612 388 3594

Offers two-year programs in renewable energy engineering, including PV.

RENEWABLE ENERGY INSTITUTE
2010 Massachusetts Avenue, N.W.
Fourth Floor
Washington, D.C. 20036

202 822 9157

RENEWABLE ENERGY PROJECTS
Box 359, R.D. 2
Ulster, Pennsylvania 18850

Brooks Eldredge-Martin
717 888 9349

PV dealer—design through installation. Renewable Energy Projects provides heat loss studies, energy use surveys, helps to develop conservation plans, and performs site studies.

RENEWABLE POWER CORP. (RPC)
460 Lincoln Center Drive
Foster City, California 94404

James Caswell
415 571 7716

PV power plant development. Private utility. Currently developing two plants utilizing hybrid PV/water heating arrays.

RES PHOTOVOLTAIC ENGINEERING INC.
P.O. Box 3084
Scottsdale, Arizona 85257

Walter O'Neill
602 829 9514

RES is a consortium of high technology professionals experienced in the engineering, design and installation of energy-conserving photovoltaic systems. RES provides total project design with specific services in: photovoltaic feasibility studies, resource and problem analysis, economic practicability studies, computer modeling, systems analysis, design and engineering services—electrical, structural, mechanical and civil, components testing, construction, site preparation, transportation of components, erection of structures, turnkey practicability studies, plus R&D in large commercial, residential or remote PV uses.

REV MANAGEMENT COMPANY INC.
P.O. Box 104
42 Grove Street
Peterborough, New Hampshire 03458

Kurt Bleicken, President

PV investment consultant.

RHO SIGMA DIV., WATSCO, INC.
1800 W. Fourth Avenue
Hialeah, Florida 33010

Hans Kolster, Sales Manager
305 885 1911

Manufacturer of solar controls and energy monitoring instrumentation including: Btu meters, temperature sensors, flowmeters, strip chart recorders, PV pyranometers, differential thermostats, simulators, solar control testers, system controllers, and series charge regulators for PV systems.

RISING SUN ENTERPRISES
26 Atlantic Avenue
Bar Harbor, Maine 04609

Photovoltaics supplier and design services.

ROBBINS ENGINEERING, INC.
3461 N. Jamaica Boulevard
Lake Havasu City, Arizona 86403

Roland W. Robbins, Jr., P.E., President
602 453 1738

Manufacturer of the Sun Seeker, positive-drive, thermohydraulic sun tracking system used worldwide for aiming concentrating collectors and photovoltaic arrays. Aiming accuracy within 2 degrees; drive force capability in excess of 300 pounds; maintains aim in winds in excess of 80 mph; no maintenance or field adjustments required; totally sealed unit with one moving part; all components corrosion-resistant and impervious to ultraviolet radiation. Requires no computers, no batteries, no electricity.

ROCKWELL INTERNATIONAL CORPORATION
Rocky Flats Test Center
P.O. Box 464
Golden, Colorado 80401

303 441 1300

PV and wind research and development.

ROCKY MOUNTAIN SOLAR ELECTRIC
5938 S. Vale Road
Boulder, Colorado 80303

Mark McCray, Managing Director
303 494 6219

Marketing of ARCO Solar panels and balance of system components; home electric and water pumping systems; and waste heat recovery systems. Classes in D-I-Y solar; solar greenhouse and sunspace design, construction and operation; PV system design and installation. Six photovoltaic power packages available with rated power outputs from 35 to 420 watts.

RODGERS & COMPANY INC.
2615 Isleta Boulevard, S.W.
Albuquerque, New Mexico 87105

Ray Bahm/Clarence Rodgers
505 877 1030

Distributor of PV equipment; design systems up to 100 kW. Specialize in PV water pumping systems. R&D in PV system design tools. Dealer training/education program.

PAUL ROSENBERG ASSOCIATES
330 Fifth Avenue
Pelham, New York 10803

Paul Rosenberg
914 738 2266

Consulting physicists. Consultation in photovoltaics and solar energy. Planning research and development.

RURAL ELECTRIFICATION ADMINISTRATION
U.S. Department of Agriculture
Washington, D.C. 20250

Christine Newman
202 382 9457

Research and development; information dissemination.

LEONARD A. RYDELL, P.E.
601 Pinehurst Drive
Newberg, Oregon 97132

503 538 5700

Microcomputer software for residential/commercial designers in the renewable energy field.

S

SACRAMENTO MUNICIPAL UTILITY DISTRICT (SMUD)
6201 S
Sacramento, California 95817

916 452 7811

Presently involved in PV project which calls for installation of 100 megawatts of PV power over next 20 years.

SANDIA NATIONAL LABORATORIES
P.O. Box 969
Livermore, California 94550

Garry Jones/Michael Thomas
415 422 2447

Research and development; testing. Current R&D in commercialization of utility-interactive PV systems.

SAN PATRICIO SOLAR
Route 2, Box 45
Mathis, Texas 78368

Lonnie Glasscock III
512 547 2256

Supplier of small and large PV systems for hunting and fishing camps, vacation cabins, RVs and boats. Sales of all related PV accessories (controllers, inverters, batteries, appliances, tools, lighting, PV books). Developing vandal- and theft-proof PV mounting systems (for rural and/or isolated cabins).

SANYO ELECTRIC COMPANY, LTD.
18 Deihan Handori 2-chome
Moriguchi, Osaka 570
Japan

Producer of amorphous silicon cells and modules; also, battery chargers.

SCIENCE ASSOCIATES
230 E. Nassau Street
Princeton, New Jersey 08540

609 924 4470

Sensors, recorders, indicators and integrators for measuring solar radiation.

SCRIBE INTERNATIONAL
1000 S. Grand Avenue
Box 15606
Santa Ana, California 92705

714 835 6660

Microcomputer software for residential/commercial designers in the renewable energy field.

SEMIX INC.
15801 Gaither
Gaithersburg, Maryland 20877

301 948 4680

Research and development. Manufacturing, silicon cell material. Affiliate of Solarex.

SERVCO PACIFIC INC. dba SERVCO INTERNATIONAL
2111 Wilcox Lane
Honolulu, Hawaii 96819

Mike Hirokawa
808 847 3770

Hawaii State and Pacific territory distributor for ARCO Solar photovoltaic modules and ancillary equipment: ARCO Solar support structures, charge controllers, Delco Remy and Exide batteries, Best Energy Systems and Tripp-Lite inverters, REC Specialists fluorescent fixtures, and Norcold refrigerators.

SHARP CORPORATION
22-22 Nagaike-cho
Abeno-ku, Osaka 545
Japan

Flat-plate and concentrating silicon cells and modules. R&D on amorphous silicon.

SIEMANS AG
Postfach 103
D-8000 Munich 1
West Germany

Single-crystal silicon cell production. R&D on monocrystalline PV film production process and amorphous silicon.

SILICON SENSORS, INC.
Highway 18 East
Dodgeville, Wisconsin 53533

Robert L. Bachner
608 935 2707

Founded in 1960, Silicon Sensors, Inc. manufactures high-efficiency silicon solar cells, photovoltaic panel arrays and power systems, and photoelectric control cells.

SILONEX INC.
331 Cornelia Street
Plattsburgh, New York 12901

Carol A. Larsen, Sales Manager
518 561 3160

Manufacturer of photovoltaic modules with peak power of 4, 9, 18 and 32 watts.

SIMPLER SOLAR SYSTEMS
P.O. Box 12936
Tallahassee, Florida 32308

Al Simpler

PV designer, R&D. Designed PV-powered medical facility and now at work on stand-alone industrial park.

SKYHEAT ASSOCIATES
Route 2
English, Indiana 47118

Richard J. Komp, Ph.D.
812 338 3163

Skyheat is a nonprofit organization dedicated to research and education in the field of solar energy. Richard Komp, author of *Practical Photovoltaics—Electricity from Solar Cells*, has been working with solar cells for over 20 years and regularly conducts hands-on photovoltaics workshops and seminars at his Indiana research facility and on tour. Extensive system design experience. Current R&D on PV/thermal hybrids, luminescent concentrators. Manufacture of Skyheat PV module designs through SunWatt, Inc.

SLYKHOUSE ENGINEERING
3033 Madison Avenue S.E.
Grand Rapids, Michigan 49508

Thomas E. Slykhouse, P.E.
616 247 0154

Engineering Science Services: Consulting, research, product and process development, materials and product testing, calculations, concept development and evaluation, patent and new idea evaluation, technical advice and support services to management.

Fields of Interest: Polycrystalline and organic photochemical cells, solar thermal engines, photoelectrochemical fuel production, electrical and thermal energy storage concepts, photovoltaic and photothermal cogeneration, waste organic recovery systems.

Advanced concept hardware and systems research. Product testing. Design of testing equipment, prototype units, thermal equipment.

SMS ENERGY CORPORATION
1095 Normington Way
San Jose, California 95136

Dave Biron
408 264 7245

SMS has been involved in the research and development, design and incorporation of photovoltaic technology into marine, RV, remote home, telecommunications, agriculture, and line-tie installations. Supplier of PV systems and auxiliary equipment, controllers, etc. R&D: prototype 4-kW line-connected system. Architecture/design services; education and information dissemination.

SOLAC BUILDERS LTD.
1610 Hoffman Drive, N.E.
Albuquerque, New Mexico 87110

505 296 8778

PV and wind power production equipment; electronic controls.

SOLAMERICA CORPORATION
1514 N. Harbor City Boulevard
Melbourne, Florida 32935

William W. Winters
305 259 2144

Solar contractor. Installation of 9-watt PV panels on our solar water heating systems—300 to 500 systems (currently) per year.

SOLAR AMERICA CORPORATION
3100 N.W. 7 Avenue
Miami, Florida 33127

Barrie Mathieson, Vice President
305 635 1472

Supplier of photovoltaic equipment. PV equipment is now standard in all our solar domestic water heating systems to power our 12-volt DC circulating pumps. Design of irrigation systems and small solar-electric generators.

SOLAR-AUDIO VISUAL ELECTRONICS
Box 4869
San Francisco, California 94101

Jim Lavandier/Glen Morris
415 824 5924

At present we distribute 12-volt DC turntables, lights, stereo amplifiers, and compact stereos. Consultation on 12-volt systems.

SOLAR CELLS OF FLORIDA
1923 West Bay Drive, No. 3
Largo, Florida 33540

Bob Laws, President
813 586 4613

PV distributor and factory representative for Solarex.

SOLAR CELLS LTD.
3327 D Mainway
P.O. Box 1025
Burlington, Ontario L7P 3S9
Canada

416 335 4713

PV module manufacturer.

SOLAR CENTER
10018 Cortez Road, West
Bradenton, Florida 33507

Kirk Maust
813 792 0391 or 813 792 8223

ARCO Solar dealer; contractor; retail sales.

SOLARCON, INC.
607 Church
Ann Arbor, Michigan 48104

Dr. Roderich W. Graeff
313 769 6588

Microcomputer programs for solar energy calculations. Software packages include Passolar, F-Chart, F-Load, PV F-Chart, Heat Loss, Glass Stress, Window Heat Transfer.

THE SOLAR CONNECTION
807 North Boulevard, West
Leesburg, Florida 32748

Henri P. Couture
904 787 4334

Licensed solar contractor; supplier of PV equipment.

SOLAR DESIGN ASSOCIATES, INC.
Conant Road
Lincoln, Massachusetts 01773

Steven J. Strong
617 259 0350

Architects and engineers specializing in the design and engineering of energy-efficient and energy-independent solar residences. Designed the country's first photovoltaic residence in Carlisle, Massachusetts, and six other PV homes. Engineering expertise for PV systems design for any application—both large- and small-scale, utility-interactive and stand-alone.

SOLAR DESIGN CONSULTANTS, INC.
14213 Banbury Way
Tampa, Florida 33624

Frank Arenas, President

Designer of PV homes. Solara II—1.2 kW PV array with reflector system for all electricity.

SOLAR DESIGNS
6010 Duclay Road
Jacksonville, Florida 32210

Nelson Hellmuth
904 778 7540 or 904 772 9776

Engineering consulting on all aspects of solar energy and energy conservation. Professional engineer specializing in renewable energy systems.

SOLAR ELECTRICAL SYSTEMS
9742 Cactus Avenue
Chatsworth, California 91311

Greg Johanson
213 998 6568

Design, sales and installation of PV equipment. DC, AC and grid systems. Specializing in remote homes and self-sufficient living at a comfortable level. Catalog and information available. R&D in prototype systems. Soon to be developing track homes on the West Coast. Presently involved in PV-powered speed record with Joel Davidson with the fastest PV-powered car without batteries to date (40 mph+).

THE SOLAR ELECTRIC CO. INC.
P.O. Box 7037
Honolulu, Hawaii 96821

Ronald C. Richmond, President
808 373 9119

Supplier of PV systems and components. Design of complete stand-alone and line-tied systems for residential, commercial, industrial, marine, communications, agricultural and institutional applications. Individual components from modules and power conditioning equipment to DC products and appliances. Also, consulting, design, project management, installation and service.

SOLAR ELECTRIC DEVELOPMENT INC.
38 Rushing Wind
Irvine, California 92714

Jack A. Bertch
714 552 9733

Solar Electric Development Inc. is a research and development organization, currently working on solar electric generators and hot water/steam systems, aimed at the rural, industrial and commercial markets. Manufacture of 50- and 259-kW solar electric generators will commence in 1984. Also manufactures and supplies parabolic dish, high-temperature solar concentrators and collectors.

SOLAR ELECTRIC DEVICES INC.
49 West Court Street
Doylestown, Pennsylvania 18901

Charles Phillips/James Thoresen
215 348 7829 or 1 800 341 4026

Supplier of small PV systems for residential use. Design of PV and energy conservation systems and products.

SOLAR ELECTRIC ENGINEERING INC.
6140 Sebastopol Road
Sebastopol, California 95472

Gary Starr/Ed Haynes
707 823 2588

SEE designs, develops, assembles, manufactures and markets various products in the alternative energy field, with special emphasis on photovoltaics. SEE also engages in the design and construction of photovoltaic power systems (PV farms) through third-party financing, for production and sales of electric power to private or corporate users; or to public utilities under federal regulations (PURPA).

SEE is engaged in PV activities ranging from educational toys and fans to complex pumping installations for domestic and foreign use.

SOLAR ELECTRIC MFG. CO.
Div., Frank E. Thompson
Box 248
Paso Robles, California 93447

Harold A. Dawson, Vice President – Production
805 239 0448

Manufacturer of PV modules to 66 watts. Supplies complete package units to power deep-well water pumps (down to 1000 feet) including auxiliary diesel-powered units to supplement electrical generation during winter months in northern territories. 100%

automatic. Firm is engaged in export sales to South America and Australia. Research and development.

SOLAR ELECTRIC POWER COMPANY
P.O. Box 653
Lincoln, Massachusetts 01773

Steven J. Strong
617 259 9426

Northeast distributor for Mobil Solar Energy Corporation. Components and turnkey systems.

SOLAR ELECTRIC SPECIALTIES CO.
P.O. Box 537
Willits, California 95490

Phil Haas/Paul McClusky
707 459 9496

SES is a full-line distributor of photovoltaic power system components featuring ARCO Solar photovoltaic modules. Other products include power inverters, deep-cycle and shallow-cycle batteries, battery chargers, engine generator sets, photovoltaic charge controllers, photovoltaic mounting structures, wire, switching and load distribution equipment and low-voltage DC lighting and appliances.

SES maintains an applications department to provide computerized custom system design assistance to dealers, contractors and engineering firms. Applications include remote housing power systems, marine power, remote communications systems, water delivery systems, railroad signaling and cathode protection. Books, publications and brochures available on PV systems.

SOLAR ELECTRIC SYSTEMS
Box 1562-A
Cave Creek, Arizona 85331

Noel Kirkby
602 488 3708

Solar Electric Systems specializes in systems consisting of usually one to four solar panels for recreational vehicle installations. We offer a complete installation package called the RV POWERPAC. The compact control panel and regulator is ideally suited for 12-volt RV system monitoring.

Aside from PV equipment, we offer a planning booklet with custom designing worksheet and resource/book list to assist the consumer in making a wise choice for a solar battery charging system. Our newsletter, *Solar Electric Update*, is published two times per year to cover up-to-the-minute PV information for RV customers and solar users.

SOLAR ELECTRIC SYSTEMS
Div. World Efficient Energy Systems
3661 63rd Avenue, North
Pinellas Park, Florida 33565

A. Michael Rankin, President
813 526 0793

We distribute ARCO Solar PV modules, REC 12-volt lights, Dytek inverters, Morning Star collectors, Norcold 12-volt refrigerator/freezers, and PV products from Solec and Specialty Concepts. We also integrate other manufacturers' products to make the Sol-lectric solar hot water systems, the first systems approved by the Florida Solar Energy Center.

Design emphasis is on remote home power and water pumping systems. Technical assistance and sizing for all PV applications.

SOLAR ELECTRIC SYSTEMS (OF NEW MEXICO) INC.
2700 Espanola, N.E.
Albuquerque, New Mexico 87110

Steve Verchinski
505 881 4765

Manufacturer of MAX power tracker and controllers. Distributor for Solec and Jacuzzi products. Design of residential PV systems, water delivery systems, telecommunications and special applications, 10–20 kW+. R&D in PV instrumentation, datalogging, utility-interactive systems, and hybrids. Also, export C.I.T.—remote home, water delivery, telecommunications, and navigation.

Educational courses and seminars offered: Solar Electric Water Pumping for Federal Land Management Personnel; Solar Electricity and Rural Health Services for BIA and Public Health Service Personnel; PV Home Seminar—Design and Implementation of Alternate Energy Living; and Government Alternate Power Seminar—Concept, Design, and Materials Specifications Writing for Photovoltaics.

SOLAR ELECTRONICS INTERNATIONAL
156 Drakes Lane
P.O. Box 39
Summertown, Tennessee 38483

Richard McKinney/Steve Skinner
615 964 2222

Solar Electronics International is a photovoltaic/wind energy systems house providing design, sales, and installation services. We sell a complete line of PV/wind components and equipment, including Solarex, ARCO, Jacobs, Bergey, Exide, and others. We can provide turnkey installation from small home units to large power systems here and abroad.

SOLAR ENERGIES OF CALIFORNIA
11421 Woodside Avenue
Santee, California 92071

Control packages, PV modules.

SOLAR ENERGY CORP.
Box 3065
Princeton, New Jersey 08540

609 924 1879

Microcomputer software for residential/commercial designers in the renewable energy field.

SOLAR ENERGY INDUSTRIES ASSOCIATION
1156 15th Street, N.W.
Washington, D.C. 20005

David Gorin
202 293 2981

Membership organization representing solar equipment manufacturers, distributors and suppliers.

SOLAR ENERGY INDUSTRIES ASSOCIATION
PV Division
P.O. Box 4400
Woodland Hills, California 91365

SOLAR ENERGY INSTITUTE OF NORTH AMERICA
1110 Sixth Street, N.W.
Washington, D.C. 20001

202 289 4411

SOLAR ENERGY INTELLIGENCE REPORT
951 Pershing Drive
Silver Spring, Maryland 20910

Allen Frank, Editor
301 587 6300

Weekly newsletter examines all solar technologies, including PV. Covers new products, industrial developments and activities worldwide, legislation and governmental events, financial matters, etc. Weekly calendar of events.

SOLAR ENERGY RESEARCH INSTITUTE (SERI)
1617 Cole Boulevard
Golden, Colorado 80401

Edward Sabisky
303 231 1236

The center for government PV research. Long-term, high-risk, advanced PV cell research and development. Emphasis on low-cost solar cells. Research activities, inhouse and under subcontract, include the areas of devices and measurements, amorphous thin films, advanced high-efficiency cells, polycrystalline silicon, and polycrystalline thin films, photoelectrochemical cells and support research.

SOLAR ENGINEERING COMPANY
460 Indian Creek Drive
Cocoa Beach, Florida 32931

Charles Cromer
305 783 6020

Distributor of PV panels, controls, pumps and inverters. Design of PV pumping systems. R&D in uses of PV for DHW pumped systems and PV homeowner-utility interface.

SOLAR ENGINEERING & CONTRACTING
P.O. Box 3600
Troy, Michigan 48007

Wayne C. Johnson, Editor
313 362 3700

Bi-monthly magazine reports latest developments in photovoltaic technology and business.

SOLAR ENVIRONMENTAL ENGINEERING CO.
2524 E. Vine Drive
Fort Collins, Colorado 80524

303 221 5166

Microcomputer software for residential/commercial designers in the renewable energy field.

SOLAREX CORPORATION
1335 Piccard Drive
Rockville, Maryland 20850

Janet Roberts
301 948 0202

Solarex Corporation designs, manufactures and installs photovoltaic power systems for a wide range of applications. Solarex photovoltaic modules feature patented semicrystalline silicon solar cells, which are produced at lower cost, provide greater space efficiency, and offer a more attractive appearance than conventional single crystal solar cell modules.

Solarex is vertically integrated with complete capabilities that range from silicon feedstock refining and cell and module manufacturing to system design, installation and service. Solarex is represented worldwide by a network of distributors trained in system installation and service. Its extensive knowledge of photovoltaic systems and applications provides Solarex with the unique ability to offer some of the most advanced photovoltaic systems commercially available today.

SOLAREX ELECTRIC LTD.
18th Floor
Sincere Insurance Building
4 Hennessy Road
Hong Kong

SOLAREX PTY., LTD.
5 Bellona Avenue
Regents Park NSW 2143
Australia

SOLAREX S.A.
Zone Industrielle Sud
CH-1196 Gland
Switzerland

SOLAREX VENTURES GROUP
1301 Piccard Drive
Rockville, Maryland 20850

Bob Edgerton, General Manager
301 948 0202

SOLAR-EYE PRODUCTS INC.
1300 N.W. McNab Road
Ft. Lauderdale, Florida 33309

Wally Starr
305 974 2500

Solar-Eye Products distributes PV panels as manufactured by Solarex. We also supply auxiliary equipment.

SOLAR GENERATORS SINGAPORE PTY. LTD.
151 Lorong Chuan
Singapore 1955

Photovoltaic cell and module manufacturer.

SOLAR HIND ENERGY CO.
747 Sand Creek Drive
Carol Stream, Illinois 60188

Chiman M. Patel, President
312 690 1883

Consulting engineering for various PV systems and solar equipment distributor. Solarex products distributor.

SOLAR INITIATIVE
180 Grand Avenue, No. 900
Oakland, California 94612

Jerry Yudelson, President
415 834 1073

Solar Initiative provides business services to public and private companies, investors, utilities and government entities involved primarily with renewable energy, energy management and energy conservation technologies and enterprises. Our special capabilities include: business plans, third-party financing programs, market research, marketing programs, government relations, community energy programs, venture capital services, seminars and workshops, and product representation.

SOLAR INTERNATIONAL LTD.
124 Maryland Route 3, North
Millersville, Maryland 21108

301 987 9666

Solarex distributor; supplier of PV cells and panels, marine panels. Systems designed for all PV applications.

SOLARLITE PHOTOVOLTAICS
G-3374 W. Flushing Road
Flint, Michigan 48504

Patrick D. Parrott
313 733 8071

Manufacturer of water pumping systems, battery cabling, and mounting hardware, including the new "Track Master" mounting system. We also are a supplier of PV hardware and accessories. Also, engineering services, component design, prototype fabrication, installation. R&D in system testing. Seminar and slide presentations offered in photovoltaic process and applications.

SOLAR LOBBY
1001 Connecticut Avenue, N.W., No. 510
Washington, D.C. 20036

Scott Sklar
202 466 6350

Publish PV information to our 60,000 members in our *Suntimes*.

SOLAR MANAGEMENT CORP.
2701 St. Julian Avenue
Norfolk, Virginia 25040

Arnold Peterson, Sales Manager
804 623 0700

Install PV-powered heating, hot water and air conditioning systems.

SOLAR MID-WEST INTERNATIONAL
108 W. Main
New London, Iowa 52645

E. E. Smith
319 367 2256

Supplier of PV hardware and accessories.

SOLARMODE INTERNATIONAL CORPORATION
1535 Sixth Street, Suite 209
Santa Monica, California 90401

W. Trevor Hewson
213 450 9994

Designer/manufacturer of PV sub-arrays. R&D contractor and developer of machinery for automatic wiring and packaging. Pursuant to the completion of a pilot production line, the machinery and technology will be available on a turnkey basis and may be purchased by prospective array manufacturers. Solarmode International is presently marketing the sub-array on a limited basis in order to develop and determine the market.

SOLAR ONE
537 S. Sequoia Drive
West Palm Beach, Florida 33409

Alan T. Boman
305 684 8177

PV supplier/installer mainly involved with circulating pump systems.

SOLAR POWER CORPORATION
20 Cabot Road
Woburn, Massachusetts 01801

Jack Nelson
617 935 4600

Large-scale photovoltaic conversion projects. Design and manufacture of complete systems, modules, cells, voltage regulators, and array structures. Residential PV power generating kit, designed to provide 12 volts DC. SPC projects include: repeater stations for Australian National Railways, community PV systems in Tunisia, Sri Lanka, Saudi Arabia and Alaska. (Subsidiary of Exxon Corporation.) Recently suspended operations.

SOLAR RATING AND CERTIFICATION CORP.
1001 Connecticut Avenue, N.W., Suite 800
Washington, D.C. 20036

202 452 0078

SOLAR RAY
5025 S.E. 50th Street
Portland, Oregon 97206

Ray Tarpey
503 771 0117

PV component supplier.

SOLARSOFT, INC.
Box 124
Snowmass, Colorado 81654

303 927 4411

Microcomputer software for residential/commercial designers in the renewable energy field.

SOLAR SUN BATTERIES
2358 Cork Circle
Sacramento, California 95822

916 421 9062

Heavy-duty, deep-cycle batteries. Photovoltaic, hydro, wind storage.

SOLARTEK
R.D. No. 1, Box 255A
West Hurley, New York 12491

914 679 5366

Microcomputer software for residential/commercial designers in the renewable energy field.

SOLARTHERM INC.
1315 Apple Avenue
Silver Spring, Maryland 20910

Carl Schleicher
202 882 4000

Solartherm has the capacity to design, develop, manufacture, and market a variety of solar photovoltaic systems on a worldwide basis. Products range from solar photovoltaic refrigerators to PV toys, PV-powered boats, pumping systems, and other innovative applications of PV technology.

SOLAR USAGE NOW, INC.
420 E. Tiffin Street
Box 306
Bascom, Ohio 44809

Joseph Deahl
419 937 2226

We package several kits for the do-it-yourselfer who wishes to explore the use of photovoltaics. We also custom-make arrays to fit specific applications under contract with the purchaser. We stock and distribute a wide variety of cells and modules and match arrays to specific consumer products such as fans, nightlights and water pumps. We also do limited array testing and publishing related to the photovoltaics industry.

SOLAR UTILIZATION NEWS (SUN)
P.O. Box 3100
Estes Park, Colorado 80517

Art Anderson
303 586 5636

SUN, a monthly solar newspaper, regularly features articles on photovoltaics. Not only does SUN cover PV products, applications, services and books, but it often features full-length stories on certain aspects of the photovoltaic industry—*e.g.*, solar irrigation using photovoltaic cells.

SOLARVISION, INC.
Church Hill
Harrisville, New Hampshire 03450

Don Best
603 827 3347

SolarVision is the publisher of *Solar Age*, a monthly magazine founded in 1976, and *Solar Industry Bulletin*, a business-oriented newsletter issued every two weeks. Both publications provide regular coverage of photovoltaics, concentrating on system design, industry news, R&D efforts, and business-related developments in the field. SolarVision also publishes *Photovoltaic System Design*, an exclusive technical report on stand-alone systems and residential retrofits.

SOLAR VOLTAICS
A-6020 Innsbruck
Liebeneggstr. 15
Austria

Leonhard Muigg
0043 5222 20 781 (international)

Solar Voltaics develops, designs, assembles and tests PV and wind-electric systems, with emphasis on water pumping systems. Some components are manufactured inhouse, others supplied from leading manufacturers worldwide.

SOLARWEST ELECTRIC
232 Anacapa Street
Santa Barbara, California 93101

Rob Robinson
805 963 9667

Solarwest Electric does system design and engineering for photovoltaic systems. We are a leading distributor for ARCO Solar, Inc. and carry all balance of system components. Our packaged systems are used in a number of applications, including: remote home, telecommunications, utility, line-tie, recreation (RV and marine), and water delivery.

SOLAR WORKS
Route 2, Box 274
Santa Fe, New Mexico 87501

A. D. Paul Wilkins
505 473 1067

Manufacturer of meter boards and regulators. Supplier of PV panels and controls. Solar and wind resource list available. Conduct PV workshops.

SOLAR WORKS PHOTOVOLTAICS
Port Clyde, Maine 04855

Sandra M. Dicksen
207 372 8067

Information coordinator and consultant. Freelance writer/lecturer with articles published in *New Roots, Yankee, Maine Times, Solar Age*. Conducts solar energy workshops.

SOLAVOLT INTERNATIONAL
3646 E. Atlanta
Phoenix, Arizona 85040

Clyde Ragsdale/Ken Hankins/Pat Walton
602 231 6408

Designer, manufacturer and supplier of complete PV systems. Distributor of balance of systems components. Functional systems include lighting, water pumping, refrigeration, remote sites, village power. R&D on low-cost ribbon process. Literature available in English, French and Spanish.

SOLEC INTERNATIONAL, INC.
12533 Chadron Avenue
Hawthorne, California 90250

George McClure
213 970 0065

Design and manufacture complete PV systems. Offer cells in 3- and 4-inch size. Standard glass face and aluminum frame modules from 10 to 66 watts. Also 10- and 35-watt Solarcharger modules fabricated with materials that allow module to be flexible and to withstand direct impacts. Offer custom modules from fraction of a watt to 80 watts. Continuing R&D in cell technologies.

SOLELECTRIC COMPANY
1600 Kitchner, Suite F
Sacramento, California 95822

Michael DeAngelis/Jim Johns
916 391 2909

Primarily, we are designers and suppliers of photovoltaic systems and equipment. Distributor for Solarex Corporation. SolElectric provides equipment, design and installation skills for varied PV applications, including water pumping, remote site systems, cathodic protection of bridges and pipelines, and telecommunications systems.

SOLENERGY CORPORATION
171 Merrimac Street
Woburn, Massachusetts 01801

C. W. Clark, Vice President – Marketing
617 938 0563

Manufacturer of PV cells, modules and support structures. Custom design of PV systems (cells or modules). Supplier of all auxiliary equipment—batteries, voltage regulators, inverters. Solenergy's Woburn headquarters houses its cell manufacturing, module encapsulation, array assembly and testing areas, and the R&D facility that continues to work with new materials in photovoltaic conversion. Representative worldwide PV system installations by Solenergy include: cathodic protection, communications, navigational aids, refrigeration, residential, telemetry, village electrification, water pumping, RVs and boats.

SOLEQ CORPORATION
8969 N. Elston Avenue
Chicago, Illinois 60646

Inverter manufacturer.

SOLLOS INC.
1519 Comstock Avenue
Los Angeles, California 90024

Milo Macha
213 203 0728

Consultation for private companies and R&D in new, low-cost processes. Present research and development of a new metallization process for PV cells—based on common metals, frit-free screenable ink to substitute for silver or other noble metals. Program sponsored by Jet Propulsion Labs (DOE).

SOL-TEMP INC.
1505 42nd Avenue, Suite 4
Capitola, California 95010

Joseph Pastore
408 462 2917

Development is progressing in the "Electropane"—photovoltaic total energy solar window system—as a complete plug-in package controlled by a programmable microprocessor thermostat. Windows, wall, skylight and roof systems are planned. Modules can be fabricated to suit architectural aesthetics, exterior and interior.

SONTEK ENERGY CORPORATION
11615 Forest Central Drive
Dallas, Texas 75243

Steve Keith
214 340 2211

Sontek Energy Corporation is a manufacturer of the SunLoc series of hybrid solar panels for space heating, hot water and photovoltaic electricity. Sontek custom designs and installs complete systems as well as oversees installations. Sontek also works with heat recovery and energy management systems which can be incorporated in the overall design or independently.

SOUTH DAKOTA OFFICE OF ENERGY POLICY
State Capitol
Pierre, South Dakota 57501

Steven Wegman
605 773 3603

State office concerned with all applications of PV technology, education and information dissemination.

SOUTHERN CALIFORNIA EDISON
118 East Carrillo
Santa Barbara, California 93101

805 963 9631

Meters and purchases electricity from ARCO Solar's one-megawatt PV power plant in San Bernardino County, comprised of 108 double-axis trackers, computer-oriented to the sun, each holding 256 PV modules. Also meters Solar One, a 10-megawatt solar thermal electric generating plant near Barstow, California.

SPECIALTY CONCEPTS, INC. (SCI)
9025 Eton Avenue, Suite D
Canoga Park, California 91304

Tom Philp
213 998 5238

Manufacturer of PV system controls, SCI offers a complete line of photovoltaic battery charge regulators, monitors, and custom control systems. Regulator types include two-step series relay, solid-state switching shunt, and solid-state series. Options include load management, temperature compensation, adjustable set points, analog or digital metering, high-low voltage alarms, and waterproof enclosures.

SPECTROLAB, INC.
12484 Gladstone Avenue
Sylmar, California 91342

213 670 1515

Manufacturer of gallium arsenide cells for space applications. R&D on advanced solar cell concepts and solar simulators.

SPIRE CORPORATION
Patriots Park
Bedford, Massachusetts 01730

Reznor Orr
617 275 6000

Spire Corporation provides business, researchers, and government organizations worldwide with technology and manufacturing equipment to produce photovoltaic cells and modules. We make it possible for organizations to become part of the growing photovoltaic industry by gaining access to state-of-the-art technical services, manufacturing processes and equipment.

Research activities include advanced crystalline and polycrystalline silicon cell technology, amorphous silicon and gallium arsenide; advanced cell processing with ion implantation and electron beam annealing; high-efficiency cells (18 – 20%); and module encapsulation using electrostatic bonding. Also, researching a unique process to grow a thin film of gallium arsenide directly on a silicon substrate to double PV cell efficiency. Research underway on advanced multijunction amorphous silicon alloy solar cells.

SSI
2107 East 5th Street
Tempe, Arizona 85281

Robert Arganbright
602 829 1903

SSI is devoted to supplying the public premium photovoltaic products, providing qualified photovoltaic engineering, and featuring demanding service. Designer and provider of PV packages for residential and remote applications.

STANDARD OIL COMPANY OF OHIO (SOHIO)
Chemical and Industrial Products Division
Midland Building
Cleveland, Ohio 44115

216 575 4141

Partner with Energy Conversion Devices in Sovonics Solar Systems to produce amorphous cells.

STANFORD UNIVERSITY
Department of Materials Science and Engineering
Stanford, California 94305

Richard H. Bube, Chairman
415 497 2534

My group is engaged in research on photoelectronic properties, materials and devices, particularly those involved in photovoltaics. The research is of both a basic and an applied nature with support from the Department of Energy through their Basic Energy Sciences Division and through the Solar Energy Research Institute. At the present time we are investigating the properties of cadmium telluride as a photovoltaic absorber material, using both single-crystal and thin-film material. Thin films of cadmium telluride are deposited by close-spaced vapor transport and by hot-wall vacuum evaporation. We are or have been also engaged in photovoltaic research involving indium phosphide, zinc phosphide, and cuprous sulfide.

Annual graduate-level course: Photovoltaic Solar Energy Conversion. Academic Press has published our book, *Fundamentals of Solar Cells: Photovoltaic Solar Energy Conversion*, by A. L. Fahrenbruch and R. H. Bube.

STERN RESEARCH CORPORATION
4444 Orcutt Road
San Luis Obispo, California 93401

805 543 4444

Manufacturer and supplier of balance-of-systems for PV.

STOVEMAN DIVERSIFIED ENERGY PRODUCTS
Route 100 Eagle
Uwchland, Pennsylvania 19480

Will Hartzell
215 458 8011

Supplier of ARCO Solar modules, BEST inverters, Gould and Exide batteries, low-voltage lighting, fans, appliances. PV design and installation services.

STRATEGIC MARKETING INC.
254 Park Avenue South
New York, New York 10010

Christopher R. Gadomski
212 475 8962

Strategic Marketing has provided marketing research services for U.S. photovoltaics firms. It has also represented the interests of a

U.S. photovoltaic firm in international markets, and it has introduced investment bankers to photovoltaic firms looking for capital.

Mr. Gadomski is co-author of *The Solar Investor's Handbook*, contributing the chapters on investing in photovoltaics.

STRATEGIES UNLIMITED
201 San Antonio Road, Suite 205
Mountain View, California 94040

William J. Murray/Robert O. Johnson
415 941 3438

Strategies Unlimited is a technology-based market research firm, providing a broad range of services to the alternative energy industries. Multi-client subscription services are provided in photovoltaics, wind, and energy management systems. These programs describe the developing markets, the technologies, the application economics and the industry infrastructure. Additional custom services are provided in solar-thermal, genetic engineering, and other technologies having application in the developing energy field. Recently completed a study on the worldwide polysilicon marketplace.

SUNAMP SYSTEMS, INC.
7702 East Gray Road
Scottsdale, Arizona 85260

Howard Barikmo, Director of Marketing
602 951 0699

Photovoltaic systems engineering and integration. Provide balance of systems designs for residential, telecommunications, water pumping, cathodic protection and remote home application. Distribute various PV modules. Manufacture all solid-state PV battery regulators for 12- and 24-volt applications, handling up to 240 watts. Manufacture control centers for residential applications that include battery regulation in sizes of 12 ampere steps, in voltages of 12, 24, 48 and 120 volts DC. Includes array and load current measurements, battery voltage and temperature using LED display

2% accurate meter. Control center may also be provided with load shedding, alarm functions and other management features. Manufacture a self-contained PV-powered 4-watt fluorescent yard light that needs only to be set in an 18-inch-deep post hole. Manufacture cathodic protection in five models with power capabilities from 100 to 2000 watts, and voltages from 12 to 120 volts DC. Manufacture a PV-powered progressing cavity positive displacement water pump that can pump to depths of 2000 feet with capacities sufficient for stock watering and domestic needs. No battery storage required. Distribute 6- and 12-volt DC deep discharge batteries that exhibit low internal resistance and excellent lifetime for use in PV storage systems.

Provide monthly three-day seminars on photovoltaics designed for prospective distributors, dealers and users of PV energy. Associated with a solar architect who designs energy-efficient pyramid homes that are capable of stand-alone electrical application or connection to the grid system.

SUN ARRAYED CONTRACTING, INC.
13 Villa Street
Roslyn Heights, New York 11577

Arnold Norman, President
516 484 2287

Manufacture self-regulating marine and recreational solar battery chargers. The units have a 5-year warranty to maintain the following specifications: SA 100 .4A, 14V; and SA 150 .9A, 14V. In addition, we produce solar-powered office and home decorations.

SUN-EARTH INTERFACE
1465 Dana Avenue
Palo Alto, California 94301

Albert Keicher
415 323 1691

PV system design—residential and small commercial applications. Seminars and workshops for general public—homeowners, small commercial businesses, real estate and financial suppliers. Community college and adult education program levels.

SUN FROST
Box DD
Arcata, California 95521

Larry Schlussler
707 822 9095

Manufacturer of Sun Frost solar-powered DC refrigerator for PV-powered sites. Low energy consumption, custom crafted, quiet, passive cold weather operation. Models available: 10 cubic foot and 17 cubic foot.

SUNNYSIDE SOLAR
R.D. 4, Box 295
West Brattleboro, Vermont 05301

Richard Gottlieb
802 257 1482

Sales and installation of PV and balance of systems components. Organizer of hands-on workshops. Photovoltaics—offered in fall course listings for University of Vermont Continuing Education Center for Southeast Vermont (Fall 1983). The physics of electricity, photovoltaic cells, arrays and systems.

SUNPOWER COMPANY
1396 Wessyngton Road, N.E.
Watertown, Massachusetts 02172

Rick Langhorst
404 874 4955

Dealer for ARCO Solar PV modules and auxiliary equipment.

SUNRACKS
Division, Sunsearch, Inc.
Box T
Guilford, Connecticut 06437

203 453 6591

Microcomputer software for residential/commercial designers in the renewable energy field.

SUN RESEARCH INC.
P.O. Box 210
New Durham, New Hampshire 03855

603 859 7110

Manufactures MAYDAY™ power supply equipment, including the 60+ W DC to AC sine wave converter, available for 12-, 24- and 48-volt DC input.

SUNRISE BUILDERS
Route 121
Grafton, Vermont 05146

Jim McCall/Kathleen Whalen
802 843 2311

Design and installation of PV systems for residential applications.

SUNRISE TECHNOLOGIES, INC./THE SUN SHOP
P.O. Box 506
Chapel Street
Block Island, Rhode Island 02807

Nancy Walker Greenaway
401 466 2122

Sunrise Technologies, Inc. designs and installs photovoltaic systems for homes, businesses, and boats. Its headquarters, The Sun Shop, is a PV-powered building shared with the local printer/

naturalist who runs her press on solar power. It is, as far as we know, the first PV-powered printing press in the nation.

In addition to our design and installation work, we provide an information outlet through our retail shop. There the public can see a complete, stand-alone system at work, compare modules of several manufacturers, find books and magazines about solar electricity, and ask questions about solar electricity. Mail orders for books and for PV components are welcome.

SUN RUN COMPANY
P.O. Box 547
Mountain View, California 94042

Dan McClory
415 961 8589

Sun Run Company is a group of high technology professionals experienced in the engineering, design and installation of PV systems. Some of the applications which can best use our services include: irrigation, desalination, stand-alone PV systems, hybrid PV and wind generator systems, hybrid PV and fossil fuel generator systems.

SUNSHINE POWER CO.
1018 Lancer Drive
San Jose, California 95129

408 446 2446

Microcomputer software for residential/commercial designers in the renewable energy field.

SUNTRONIC/SOLAR-ELECTRONIC
P.O. Box 60 53 44
D-2 Hamburg 60
West Germany

040 440959

Manufacturer of flexible PV panels. PV component and auxiliary equipment supplier; large PV catalog available.

SUN-UP SOLAR SECURITY
P.O. Box 367
Kent, Washington 98032

Will Wilson
206 854 8420

Technical and commercial advisory in solar energy, environmental applications. Design—new systems technology, computer search, software.

SUNWATER CONSTRUCTION
219 Van Ness Avenue
Santa Cruz, California 95060

George B. Hue
408 423 2429

Sunwater provides direct sales and installation of PV systems for residential, marine and recreational vehicle uses. We also can make available specially designed and engineered systems to meet customers' special needs. We are associated with Sol-Temp, Inc. and will be a major distributor of this innovative solar product.

SUNWATT CORPORATION
Route 2
English, Indiana 47118

Richard J. Komp, President
812 338 3163

SunWatt Corporation manufactures hybrid PV/thermal modules and marine PV modules. Full-range PV supplier. Custom design of PV modules and complete systems. R&D in silicon solar cell technology and module design.

SUNWATT INTERNATIONAL
Box 24167
Denver, Colorado 80224

Carol Harlow, President
303 758 4910

SunWatt International is involved in export and overseas licensing of photovoltaic production technology, in joint ventures.

SUNWORKS
7741 S.W. Capitol Highway
Portland, Oregon 97219

Daniel J. Merkle
503 245 5650

Contracted by Oregon Department of Energy to install a photovoltaic array consisting of 39 modules at the Williamette Mission State Park in Gervais, Oregon, April 1982. This is the largest residential photovoltaic system in the Pacific Northwest (1.5 kW).

Contracted by Bonneville Power Administration to design, provide and install the largest PV array (280 modules) in the Northwest at the Bonneville Power Administration Building in Redmond, Oregon, July 1982 (10 kW).

Also, sales, installation and leasing of water pumping systems.

SURRETTE STORAGE BATTERY CO., INC.
16 Front Street
P.O. Box 3027
Salem, Massachusetts 01970

Paul J. Arrington
617 745 4444

Surrette manufactures a complete line of 6-, 8- and 12-volt heavy-duty, deep-cycling lead-acid storage batteries for use in any voltage and amp hour capacity system. 50 years experience in marine, commercial, industrial, and alternate energy field.

T

TALMAGE ENGINEERING
Beachwood Road
P.O. Box 497A
Kennebunkport, Maine 04046

Peter Talmage
207 967 5945

Supplier of ARCO modules and complete line of accessories. System design for maximum efficiency and continuing testing of new equipment for use with PV.

TANGENT ENTERPRISES
P.O. Box 266
Del Mar, California 92014

619 481 2444

Module packaging.

TECHNICAL INFORMATION CENTER
U.S. Department of Energy
P.O. Box 62
Oak Ridge, Tennessee 37830

615 576 5454

TELEFUNKEN
P.O. Box 1109
D-7100 Heilbronn
West Germany

Developer of polycrystalline silicon cells. Largest European PV cell and module manufacturer.

TENNESSEE VALLEY AUTHORITY (TVA)
Solar Applications Branch
2117 Andy Holt Avenue
Knoxville, Tennessee 37902

615 522 7026

Contracting with Entech for prototype PV power plant that TVA will use for test and operational experience in their test facility near Chattanooga.

TENSEN CO., INC.
304 S.E. Second
Portland, Oregon 97214

503 239 5922

Manufacturer of 12-volt DC chainsaws.

TERRALAB ENGINEERS
3585 Via Terra
Salt Lake City, Utah 84115

801 262 0094

Testing agency. Complete technical, labeling, follow-up services, including fire and flammability, efficiency, hail testing, electrical, fluid and mechanical testing.

TESLACO
490 South Rosemead Boulevard, Suite 6
Pasadena, California 91107

Slobodan Cuk, President
213 795 1699

TESLAco manufactures a 4-kW 20-kHz utility interactive photovoltaic inverter. Educational division offers power electronics courses.

UNIVERSITY OF TEXAS
Laboratory for Electrochemical Research
Austin, Texas 78712

Allen J. Bard, Director
512 471 3761

Photoelectrochemical research on semiconductor electrodes for solar energy conversion.

TEXAS ELECTRONICS, INC.
5529 Redfield Street
Dallas, Texas 75209

214 631 2490

Manufacturer of modular meteorological systems, signal processors, wind speed and direction sensors, air temperature, humidity and barometric pressure sensors, electric rain gauge transmitters, pyranometers, and photovoltaic solar radiation sensors (total sun and sky).

TEXAS INSTRUMENTS, INC.
P.O. Box 5012
Dallas, Texas 75222

214 238 2334

R&D to incorporate photovoltaic spheres, hydrogen production and storage, a fuel cell and heat exchanger in modular, integrated power system.

THERMAL SPECIALTIES
420 S.E. Main Street
Roseburg, Oregon 97470

Bill Mann
503 673 0585

PV supplier—components and systems. Technical assistance and economic analysis.

THICK FILM SYSTEMS
Division, Ferro Corporation
324 Palm Avenue
Santa Barbara, California 93101

Jason Provance
805 963 7757

Thick Film Systems produces a number of thick film conductor pastes for solar cells. Conductrox® 3347 silver is used for front collector grid metallization and Conductrox 5540 aluminum and 3398 aluminum-doped silver are used for back surface ground plane on silicon wafer-type solar devices.

THOMAS & BETTS CORPORATION
920 Route 202
Raritan, New Jersey 08869

Bernie Gudaitis, Manager – Market Development
201 685 1600

Thomas & Betts, an electrical supply manufacturing firm, in cooperation with Chronar is developing a photovoltaic raceway interconnect system, designed to meet the present and future interconnect needs of the solar electricity market.

TIDELAND ENERGY PTY. LTD.
P.O. Box 519
Brookvale, N.S.W. 2100
Australia

John Roydhouse, Marketing Manager
02 938 5111

Manufacturer of single-crystal, high-efficiency, surface textured N+/P silicon solar cells, 100 mm or 76 mm diameter. Aluminum or silver backed. Laser scribing to shape or size possible. Front contact custom design readily accommodated.

Manufacturer of high-density harsh environment twin glass solar modules—modules individually designed to suit special requirements. Manufacturer of complete solar arrays for communication, navigation, industrial and domestic applications.

Manufacturer and distributor of complete household systems, water pumping systems, and photovoltaic appliances and accessories.

TIDELAND SIGNAL CORP.
4310 Director's Row
P.O. Box 52430
Houston, Texas 77052

713 681 6101

TIOS SOLAR DESIGN & ENGINEERING
150 South 600 East
Salt Lake City, Utah 84058

801 363 3661

Architecture and engineering, residential structures incorporating PV systems.

TOPAZ, POWERMARK DIV.
3855 Ruffin Road
San Diego, California 92123

R. Baietto

Manufacturer of inverters, converters, regulators.

TOSHIBA CORPORATION
1 Komukai, Toshiba-cho
Saiwai-ku, Kawasaki 210
Japan

Silicon ribbon research and development.

TOTAL ENVIRONMENTAL ACTION INC.
Church Hill
Harrisville, New Hampshire 03450

608 827 3375

Architecture/engineering incorporating photovoltaic systems.

TRINITY UNIVERSITY
Physics Department
San Antonio, Texas 78284

Fred Loxsom
512 736 7421

Trinity University offers a course in photovoltaic system design as part of its graduate program in applied solar energy.

TRIPP-LITE
500 N. Orleans
Chicago, Illinois 60610

312 329 1777

Manufacturer of inverters.

TRISOLAR CORPORATION
10 DeAngelo Drive
Bedford, Massachusetts 01730

Theodore Osgood
617 275 1200

TriSolar is a PV systems company with design, engineering and installation services. Also, design and manufacture control equipment for photovoltaic systems. Emphasis in residential, village, water pumping, navigation and communications systems in third world countries.

TROJAN BATTERIES COMPANY
12380 Clark Street
Santa Fe Springs, California 90670

Manufacturer of deep-cycle batteries for use with photovoltaic systems.

TROPICAL SYSTEMS
P.O. Box 5278 UOG Station
Mangilao, Guam 96913

Bruce Best
Guam 789 2304

Active in PV and wind applications for tropical island and marine use. Design and manufacture of control panels, photo-wind hybrids; supplier of gen-set hybrids. R&D in tropical energy systems, OUTPUT, weathering, aquaculture applications. Conduct college-level alternative energy seminars. Publish catalog and newsletter and offer free design consultation to nonprofit organizations.

TROUT CREEK POWER CO.
694 Trout Creek Road
Republic, Washington 99166

Jerry Graser
509 775 3443

Manufacturer of control/monitor panel. Supply PV components, batteries and accessories. Conduct introductory PV workshops.

U

EARL D. UNGER
815 – 817 New York Avenue
Martinsburg, West Virginia 25401

304 267 2673

Manufacturer of a photovoltaic concentrator/solar collector mounted on suntracker.

UNITED ENERGY CORP., dba UEC
420 Lincoln Centre Drive
Foster City, California 94404

Kristi Mauch, Director of Communications
415 570 5011

Manufacture and sale of photovoltaic/thermal modules incorporating point-focus Fresnel lens optics and actively cooled silicon concentration devices. UEC is also a developer of integrated renewable energy farms which provide electrical and thermal energy, and produce ethanol and protein from biomass, as well as producing food and potable water. The "solar farm" at Barstow, California, will eventually consist of 61 ponds with floating PV arrays.

UNITED SOLAR ENERGY CORPORATION
79 Madison Avenue
New York, New York 10016

Charles Hope, President/Founder
212 684 0335

Design and construction of combination hybrid system energy parks, integrated systems to provide industrial outputs. Constructing a breeder farm on Long Island, for the light assembly of PV modules. Install and manufacture wind rotors and PV panels for residential and commercial use. Research activities include the feasibility of breeder farms, incorporating PV wind farms; solar thermal applications.

URBAN OPTIONS
135 Linden Street
East Lansing, Michigan 48823

517 351 3757

Urban Options is a nonprofit organization which provides mid-Michigan residents with information and experience related to energy conservation and appropriate technology.

ISHRAT H. USMANI
425 East 58th Street, No. 23G
New York, New York 10022

212 754 8432 or 212 751 7069

Senior energy advisor to the United Nations Department of Technical Cooperation for Development. (Bi Jilong, Undersecretary General to the United Nations, 1 UN Plaza, New York, New York 10017.) Advises the developing countries of Asia, Africa and Latin America on the use of photovoltaic cells for rural electrification and special end uses such as water pumping, desalination, ice-making, etc.

UTL CORPORATION
4500 West Mockingbird Lane
Dallas, Texas 75209

214 350 7601

R&D on gallium arsenide cells and concentrating arrays.

V

VANNER INC.
745 Harrison Drive
Columbus, Ohio 43204

Rick Wisman
614 272 6263

Manufacturer of inverters, electronic components. Patented "Ultra-light" inverter provides 3000 watts, 120 volts AC, and weighs in at 52 pounds; 1500 watt, 24 pounds. Low battery monitor automatically transfers load from inverter to line/auxiliary power. When batteries regain power from PV or auxiliary charging source, load reverts back to inverter. "Ultra-light" weight allows for concealing inverter in ceiling or other limited-space applications.

VARIAN ASSOCIATES
611 Hansen Way
Palo Alto, California 94303

415 493 4000

Advanced PV cell research. R&D on gallium arsenide cells and concentrating arrays.

VITA, INC. (Volunteers in Technical Assistance)
1815 N. Lynn Street, Suite 200
Arlington, Virginia 22209

John R. Lippert, Technical Advisor
703 276 1800

VITA provides to international aid agencies, governments, organizations and individuals in third world countries: information on photovoltaics; advice on system design and sizing for various PV applications; names and addresses of manufacturers and suppliers of PVs and related DC equipment, including overseas distributors where applicable; assistance in establishing joint ventures, licensing arrangements, and the like, to set up manufacturing plants in developing countries for PV cells, modules, and related PV equipment; on-site consultancies for the installation and operation of PV systems; testing of PV prototypes or new products under harsh field conditions in developing countries.

VOLNY ENGINEERING
285 S.E. Scott
Bend, Oregon 97702

Norbert Volny
503 382 6286

Alternative energy systems analysis and development; passive and active solar systems design; value engineering; earth sheltered structure engineering; structural engineering; energy audits; failure analysis, investigations and "forensic" engineering; thermal-structure interaction analysis; computer programming for analysis and instrumentation; engineering for arctic conditions.

VOLT ENERGY MANAGEMENT SERVICE
0434 Southwest Iowa
Portland, Oregon 97201

503 244 3613

Microcomputer software for residential/commercial designers in the renewable energy field.

WACKER SILTRONIC CORPORATION
7200 N.W. Front Avenue
Portland, Oregon 92703

503 243 2020

R&D and production of polycrystalline silicon; semicrystal cell production.

WARMRAYS INC.
Frost Road
Washington, Massachusetts 01223

Edward E. Bond
413 499 2497

Warmrays began working in the fields of energy conservation and renewable energy resources in 1976 and have developed, designed and installed photovoltaic systems for both residential and commercial use in the Northeast. They have designed and installed a 0.5 kW, 12-volt DC, battery storage, gen-set backup system on a remote residence. It was completed in March 1981 and featured in *New Shelter*.

WASHINGTON ENERGY EXTENSION SERVICE
Seattle University
Seattle, Washington 98122

Stan Price
206 626 6225

Free public education programs, with classes on photovoltaics held every other month.

UNIVERSITY OF WASHINGTON
Joint Center for Graduate Study
Seattle, Washington 98195

206 543 2100

WASHINGTON UNIVERSITY
Box 1106
St. Louis, Missouri 63130

Joseph M. Cohen
314 889 5455

Simplified reliability studies of residential PV systems; microcomputer model.

Advanced PV cell research and development.

WEBB ELECTRONICS INC.
272 A S. Monaco Parkway
Denver, Colorado 80224

W. Webb, Jr., President
303 321 0669

Solar control systems and test equipment including the WSR Regucharger, a battery charger–regulator designed specifically for use with PV systems; PV systems and accessories.

WELL-BEING PRODUCTIONS
P.O. Box 757
Rough & Ready, California 95975

David Copperfield/Barbara Copperfield
916 432 2426

Authors of *The Electrical Independence Booklet Series*. Written for the do-it-yourselfer, the series gives step-by-step instructions for installing and using photovoltaic panels: how to mount, wire, regulate and monitor the panels, how to convert turntables and washing machines to 12-volt DC, how to create a low-cost back-up system, and more. Special reports for specialized related information researched and prepared on request. Complete list available.

WESTERN NEW ENGLAND SOLAR
118 Maple Street
Holyoke, Massachusetts 01040

Tim Pafik
413 533 7750

Distributor of ARCO photovoltaic modules for remote home, recreational vehicle, navigational, telecommunication, and stand-alone lighting systems. We also carry a variety of electronic charge controllers and system monitors for PV systems as well as batteries and electrical installation accessories. Included in the accessory product line are low-voltage lights such as halogen, fluorescent, and low-pressure sodium. DC-powered refrigerators, motors, fans, inverters, and pumping systems for crop and animal watering are also available. Design services are available, and all equipment is discounted. Also, testing of PV products for quality and function.

WESTERN SOLAR PRODUCTS
1202 Hunter Avenue
Santa Ana, California 92705

714 558 7950

PV and wind equipment supplier for residential and commercial applications. Services include engineering, consulting and leasing.

WESTERN WARES
Box C
Norwood, Colorado 81423

303 327 4898

Microcomputer software for residential/commercial designers in the renewable energy field.

WESTINGHOUSE ELECTRIC CORPORATION
Advanced Energy Systems Division
P.O. Box 10864
Pittsburgh, Pennsylvania 15236

412 892 5600 or 1 800 821 7700

R&D in high-efficiency silicon solar cells. Production of single-crystal silicon ribbons by the dendritic web process.

J. C. WHITNEY
1917–19 Archer Avenue
P.O. Box 8410
Chicago, Illinois 60680

Source for 12-volt DC lights, pumps, meters, cables, appliances.

WILMORE ELECTRONICS CO., INC.
P.O. Box 2973
West Durham Station
Durham, North Carolina 27705

919 489 3318

Manufacturer of inverters, converters, UPS.

WINCO DIV. DYNA TECHNOLOGY INC.
7850 Metro Parkway
Minneapolis, Minnesota 55420

L. Attema
612 853 8400

Manufacturer of generators, wind generators, inverters.

WINDFARM MUSEUM, INC.
R.F.D. Box 86
Vineyard Haven, Massachusetts 02568

Peter Tailer
617 693 3658

We are a museum of wind, solar, and alternative technology. We have several photovoltaic exhibits, photovoltaic electric fence for our farm animals, photovoltaic air pump for fish farm exhibit, and photovoltaic-operated TV in the windpower/solar home. We intend to expand our PV exhibits as the technology becomes more affordable. Eventually the solar home will be both wind- and PV-powered.

WINDLIGHT WORKSHOP
Solar-Electric Products
P.O. Box 6015
Santa Fe, New Mexico 87502

Windy Dankoff
505 471 9299

WindLight Workshop has designed photovoltaic and wind-electric installations since 1977. Mail-order supplier: PV system components, DC supplies, tools, pumps and accessories. Catalog and handbook of unique supplies and information. Design services by mail: PV home and water supply systems. Manufacturer (wholesale/retail): "SLOWPUMP" high-efficiency, low-volume shallow well and pressure-booster pumps. DC washing machine and sewing machine conversion kits. 12-volt grain mills. Conduct local workshops and courses.

WINDWORKS, INC.
Route 3, Box 44A
Mukwonago, Wisconsin 53149

Rosanne Belair
414 363 4088

Manufacturing of power conditioning equipment for PV use. Design and build systems as large as 1 megawatt in capacity.

WISCONSIN DIVISION OF STATE ENERGY
P.O. Box 7868
Madison, Wisconsin 53707

Barbara Samuel
608 266 8871

Public information on all renewable resources to state residents and others seeking information. Also serve as an information clearinghouse and make referrals when possible. Publish energy newsletter and fact sheets.

UNIVERSITY OF WISCONSIN
Solar Energy Laboratory
1500 Johnson Drive
Madison, Wisconsin 53706

608 263 1589

Microcomputer software for residential/commercial designers in the renewable energy field.

THE WORKBOOK
P.O. Box 4524
Albuquerque, New Mexico 87106

Julie Jacoby
505 262 1862

The Workbook is a sourcebook of information on environmental, social and consumer issues for individuals and groups who want greater control over their lives. The *Self-Reliance Journal*, a 16-page section included in each issue, deals with a single topic of consumer interest, providing important basic information in handy form on a single subject.

WORLDWATCH INSTITUTE
1776 Massachusetts Avenue, N.W.
Washington, D.C. 20036

David McGregor
202 452 1999

Worldwatch Institute is an independent, nonprofit research organization. It was established to alert policy makers and the general public to emerging global trends in the availability and management of resources—both human and natural. The research program is designed to fill the gap left by traditional analyses in today's rapidly changing and interdependent world.

WYLE LABORATORIES
7800 Governor's Drive, West
Huntsville, Alabama 35807

Research and development; testing.

X

XENON CORPORATION
66 Industrial Way
Wilmington, Massachusetts 01887

Solar simulators for testing PV panels and cells.

Z

PETER G. ZAMBAS
21800 Marylee Street, Unit No. 50
Woodland Hills, California 91367

213 348 5254

Management consulting. General consulting for PV. Past experience includes Vice President and General Manager, ARCO Solar; mergers and acquisitions; organizational development; Chairman of PV Division of SEIA (1979).

C. H. ZATSICK
255 Cole Street, N.E.
Marietta, Georgia 30060

404 424 2506

Teach a career development course at Lockheed-Georgia for engineers—solar introduction including PV. Also, a continuing education course at Kennesaw Junior College in Marietta, Georgia. Introductory course in solar, including PV. Information dissemination through the Georgia Solar Coalition.

ZIPP & ZONEN
390 Central Avenue
Bohemia, New York 11716

Manufacturer of net radiometers, pyranometers and pyroheliometers.

ZOMEWORKS CORPORATION
P.O. Box 25805
Albuquerque, New Mexico 87125

Tim Wolfe, Sales
505 242 5354

Zomeworks manufactures and sells passive solar trackers for photovoltaic arrays. Also sells 12-volt circulating pumps for domestic water heating systems.

Contact Name Index

Ahline, Fred — March Mfg. Inc. 73
Alldrin, Chuck — Energy Alternatives 37
Anderson, Art — *Solar Utilization News* 126
Annan, R. H. — Department of Energy 30
Arenas, Frank — Solar Design Consultants, Inc. 112
Arganbright, Robert — *Photovoltaics* 91; SSI 133
Arrington, Paul J. — Surrette Storage Battery Co., Inc. 141
Atkins, Brent R. — Atlantic Solar Power, Inc. 14
Attema, L. — Winco 156

Bachner, Robert L. — Silicon Sensors, Inc. 107
Bahm, Ray — Rodgers & Company Inc. 103
Baietto, R. — Topaz 146
Bard, Allen J. — University of Texas 144
Barikmo, Howard — Sunamp Systems, Inc. 135
Barse, William — Monegon Ltd. 80
Beckman, Sylvia — F-Chart Software 44
Bay, Richard — Maine Office of Energy Resources 72
Becker, Bill 53
Beckman, W. A. — F-Chart Software 44
Belair, Rosanne — Windworks, Inc. 157
Benjamin, Arthur D. — ADB Engineers, Inc. 2
Benson, C. M. — Mississippi County Community College 78
Bertch, Jack A. — Solar Electric Development Inc. 113
Bertoia, Val — Bertoia Studio, Ltd. 17
Best, Bruce — Tropical Systems 148
Best, Don — SolarVision, Inc. 127
Bifano, William — NASA Lewis Research Center 81
Bingler, Douglas — Hartell 54
Bingley, Don — Acheval Wind Electronics 1
Birdwell, Gary — Georgia Power Co. 49
Biron, Dave — SMS Energy Corporation 109

Blake, John III — Heliotrope General, Inc. 56
Bleicken, Kurt — REV Management Company Inc. 102
Boltz, Benjamin D. — Insolation Solar 63
Boman, Alan T. — Solar One 124
Bond, Edward E. — Warmrays Inc. 153
Borgo, Pete — Meridian Corporation 77
Bowen, Duane H. — Duane's Solar Energy Co. 31
Brandt, Richard — New Jersey Department of Energy 84
Brown, Eugene — PV Power Corp. 97
Bru, Malcolm — Aldermaston Sales 4
Brunt, Linda — New York State Energy Office 85
Bube, Richard H. — Stanford University 133
Buckles, D. — The Energy Shop, Inc. 41
Burcham, R. Frank, Jr. — Public Service Company of New Mexico 95

Campbell, Ed — *Alternate Energy Transportation Newsletter* 5
Cannon, Don — A.H.S. Energy Supply 3
Capone, John — CW Electronic Sales Co. 28
Carbone, Robert C. — Carbone Investment Management Corporation 22
Carpenter, Linda — Florida Solar Energy Center 45
Carroll, Debra — Northern California Solar Energy Association 86
Caswell, James — Renewable Power Corp. 101
Charlins, B. — Charlins, Inc. 23
Chen, Wayne C. — University of Florida 45
Clark, C. W. — Solenergy Corporation 129
Clayton, Harry H. — IOTA Engineering Co. 64
Clevey, Mark — Kalamazoo Energy Office 66
Clifton, Michael — MC Solar Engineering 75
Cohen, Joseph M. — Washington University 154
Cook, Robert — American Standards Testing Bureau, Inc. 7
Cook, Stephen — compuSOLAR 25
Cookman, Dick — Enerdyne Solar & Wood Systems 36
Copley, Ernest — Copley Energy, Inc. 26
Copperfield, Barbara — Well-Being Productions 154
Copperfield, David — Well-Being Productions 154
Cornelison, Peter — Condar Co. 25
Cosby, Ronald M. — Ball State University 16
Couture, Henri P. — The Solar Connection 111
Cramer, Patricia — REC Specialties, Inc. 99
Cromer, Charles — Solar Engineering Company 119
Cuk, Slobodan — TESLAco 143

PHOTOVOLTAICS EDITION

Cullen, Jim — Real Goods Trading Company 99
Cummins, Richard C. — Photocomm Incorporated 90
Curry, Richard — *PV Insiders Report* 96

Dankoff, Windy — WindLight Workshop 157
Davidson, Joel — 113; *PV Network News* 29; Wm. Lamb Co. 69
Davis, Kevin — CAM-LOK 21
Dawson, Harold A. — Solar Electric Mfg. Co. 114
Deahl, Joseph — Solar Usage Now, Inc. 126
DeAngelis, Michael — SolElectric Company 129
DeNapoli, Pete — Encon Photovoltaics 36
Dennis, V. — Leveleg 70
Dickson, Sandra M. — Solar Works Photovoltaics 128
Dostal, Mark D. 31

Ebert, C. J. — Gates Energy Products 48
Eckel, John — A. Y. McDonald Mfg. Co. 75
Edgerton, Bob — Solarex Ventures Group 121
Eggleton, Wes — Appropriate Energy Management 9
Egnor, Terry — Edmonds Community College 34
Eldridge-Martin, Brooks — Renewable Energy Projects 101
Ellis, Ann — Mobil Solar Energy Corporation 79
Ellis, Bill 53
Enz, Pat — Red Wing Energy Education Center 100
Epstein, Robert T. — Dynamic Solar Products Inc. 32

Fahrenbruch, A. L. — Stanford University 134
Fazzolare, Rocco — University of Arizona (SERF) 12
Feldman, Stephen — University of Pennsylvania 90
Fitzgerald, Mark C. — *Photovoltaics International* 97
Fitzgerald, Phyllis L. — Environmental Alternatives 42
Floss, S. W. — Arnold Greene Testing Laboratories, Inc. 51
Foster, Kenneth 46
Fousel, Dale — Northern California Solar Energy Association 86
Frank, Allen — *Solar Energy Intelligence Report* 119
French, P. V. — DSET Laboratories Inc. 31
Friedman, Phil — Home Energy Workshop 58
Froland, Jerome — John Fluke Mfg. Co. 46

Gadomski, Christopher R. — Strategic Marketing Inc. 134
Garcia, J. J. — Nordika Systems, Inc. 86
Garner, Iain F. — Howard Design B.V. 59

Gilbert, Robert — Qu.E.S.T. 97
Gillett, Drew A. 49
Girouard, Phyllis — Broken Plow Law Office 20
Glasscock, Lonnie, III — San Patricio Solar 105
Gliniecki, Mary Jo — Michigan Energy Administration 78
Gorin, David — Solar Energy Industries Association 118
Gottleib, Richard — Sunnyside Solar 137
Graeff, Roderich W. — Solarcon Inc. 111
Grambs, Peter — Monegon, Ltd. 80
Graser, Jerry — Trout Creek Power Co. 148
Greenaway, Nancy Walker — Sunrise Technologies, Inc. 138
Gregory, Alvin L. 52
Grgurich, Cedric — ARCO Solar, Inc. 10
Grimbaldi, C. Lawrence — AREMCO Products, Inc. 11
Grosskreutz, J. Charles — Black & Veatch 18
Gudaitis, Bernie — Thomas & Betts Corporation 145
Gumbs, Ronald W. — Gumbs Associates, Inc. 53
Gurne, David — Arctic-Kold Energy Products 11

Haas, Phil — Solar Electric Specialties Co. 115
Hagens, Bethe 53
Haley, Robert B. 54
Hamel, Robert O. — Gates Energy Products 48
Hamrin, Jan — Independent Energy Producers 62
Hanipel, G. — Gold Star Energy Saving Products Ltd. 51
Hankins, Ken — Solavolt International 128
Harlow, Carol — SunWatt International 141
Hartley, James — Communications Associates 24
Hartzell, Will — Stoveman Diversified Energy Products 134
Hartzell, William — Independent Home Energy Systems 62
Haynes, Ed — Solar Electric Engineering Inc. 114
Heins, Conrad — Jordan Energy Institute 66
Hellmuth, Nelson — Solar Designs 112
Helms, John — AFG Industries, Inc. 3
Hensley, Stephen Michael — Last Chance Homesteads 70
Herick, Roger — Alternative Energy Engineering 5
Herman, Clifford J. — Crystal Systems, Inc. 27
Herrlinger, Gary — Mr. Sun, Inc. 80
Hershey, Robert Alan — New Volt Solar Electric 85
Hewson, W. Trevor — Solarmode International Corporation 123
Hibbein, David — Power Pak 94
Hill, Jon — The Earth Store 33

Hirokawa, Mike — Servco Pacific 107
Hood, Bobbie Sue — Hood Miller Associates 58
Hope, Charles — United Solar Energy Corporation 149
Horobin, David — Energy Compliance Systems, Inc. 37
Hoyord, Richard — Flad & Associates 44
Hue, George B. — Sunwater Construction 140
Hufford, Howard — Missouri Department of Natural Resources 79

Jacoby, Julie — *The Workbook* 158
Jilong, Bi — United Nations 150
Johanson, Greg — 29; Solar Electrical Systems 112
Johns, Jim — SolElectric Company 129
Johnson, Dave — Homestead Electric 58
Johnson, Robert O. — Strategies Unlimited 135
Johnson, Wayne C. — *Solar Engineering & Contracting* 120
Jones, Garry — Sandia National Laboratories 105
Judd, Cully — Hawaiian Solar Electric 55

Katz, David — Alternative Energy Engineering, Inc. 5
Keck, William — Keck & Keck 67
Keenan, Gerry — Free Energy Systems, Inc. 47
Keicher, Albert — Sun-Earth Interface 136
Keith, Steve — Sontek Energy Corporation 131
Keller, Doug — Atlantic Solar Power, Inc. 14
Keller, William — Los Alamos National Laboratory 72
Kelly, Paul — Energy Sciences 40
Kenedi, Ron — Independent Power Company 62
Kent, Brian — Maine Office of Energy Resources 72
Kirby, John — Missouri Department of Natural Resources 79
Kirkby, Noel — Solar Electric Systems 116
Klein, S. A. — F-Chart Software 44
Klemmer, Greg — Energy Sciences 41
Kleyman, Jim — Energy Research & Design Associates 39
Koester, Robert J. — Ball State University (CERES) 16
Kolster, Hans — Rho Sigma 102
Komp, Richard J. — 53, 57; Skyheat Associates 108; SunWatt Corporation 140
Krodt, Kenneth — C&D Power Systems 22
Kvit, Steven — Long Island Solar Energy Association 71

Lamb, William 69
Landsberg, Peter T. 69

Lane, Grey — Photowatt International, Inc.   92
Lane, Richard L. — Kayex Corporation   67
Lane, Tom — Pulstar Corporation   95
Langhorst, Rick — Sunpower Company   137
Larsen, Carol A. — Silonex Inc.   108
Lavandier, Jim — Solar-Audio Visual Electronics   110
Laws, Bob — Solar Cells of Florida   110
Leavenworth, Phil — CW Electronic Sales Co.   28
Lehn, Ted P. — Eureka Design   43
Levin, Carol — NEPEPCO   83
Levy, Sheldon, L. — Black & Veatch   18
Lewandowski, Al — Long Island Solar Energy Association   71
Lintz, Gary — Edmonds Community College   34
Lippert, John R. — VITA, Inc.   151
Liverette, Jerry — GML Systems, Inc.   50
Lourenco, Larry — Photovoltaic Power Systems   91
Loxsom, Fred — Trinity University   147
Lyon, Ervin F. — American Power Conversion Corporation   6

Macha, Milo — Sollos Inc.   130
MacQueen, T. A., Jr. — University of Connecticut   26
Madigan, Steve M. — Danfoss Inc.   28
Manetas, Mike — Northcoast Solarworks   86
Mann, Billy — Thermal Specialties   144
Marier, Abby — *Alternative Sources of Energy*   5
Martin, Bryan — M. Hutton & Co.   60
Martin, Randy — Iowa Energy Policy Council   65
Mastaitis, Vicki — New York State Energy Office   85
Mathers, Kris L.   74
Mathieson, Barrie — Solar America Corporation   110
Mauch, Kristi — UEC   111
Maust, Kirk — Solar Center   111
Maycock, Paul — PV Energy Systems   96
McCall, Jim — Sunrise Builders   138
McCarney, Steve — Colorado Mountain College   24
McClory, Dan — Sun Run Company   139
McClure, George — Solec International, Inc.   129
McClusky, Paul — Solar Electric Specialties   115
McCray, Mark — Rocky Mountain Solar Electric   103
McGory, Mike — Home Energy Workshop   57
McGowan, Rick — 76; Associates in Rural Development, Inc.   13
McGregor, David — Worldwatch Institute   159

McKinney, Richard — Solar Electronics International 117
Melody, Ingrid — Florida Solar Energy Center 45
Menashian, Lena 76
Mendenhall, James — Kansas University Solar Energy Club 67
Merkle, Daniel J. — Sunworks 141
Merrigan, Tim 77
Mieger, Bob — Lane Energy Center 70
Millard, Michael G. — Kee Industrial Products Inc. 67
Miller, Larry — Energy Sciences 40
Miller, Tom — *Arkwork Review* 13
Mize, Joe H. — Oklahoma State University 88
Moore, James S. — Mueller Associates 81
Morin, Leo D. — Free Energy Options 47
Morris, Glen — Solar-Audio Visual Electronics 110
Morsell, Lee — Paradise Power Company 89
Mueller, Donald D. — Colorado Technical College 24
Muigg, Leonhard — Solar Voltaics 127
Murphy, Steve — Florida Solar Heating Systems, Inc. 46
Murray, William J. — Strategies Unlimited 135

Naff, George J. — Hughes Aircraft 59
Neill, D. Richard — University of Hawaii 55
Nelson, Jack — Solar Power Corporation 124
Newburn, Jeff — Iowa Energy Policy Council 65
Newman, Christine — Rural Electrification Administration 104
Norman, Arnold — Sun Arrayed Contracting, Inc. 136
Norton, J. H., Jr. — North Wind Power Co. 87

O'Mara, Bradley E. — BOSS 15
O'Neill, Walter — RES Photovoltaic Engineering Inc. 101
Orr, Reznor — Spire Corporation 132
Osgood, Theodore — TriSolar Corporation 147
Overeem, Mathew — IECO 61

Pafik, Tom — Western New England Solar 155
Parker, Gary — GPL Industries 51
Parker, R. L. — Martin Marietta Denver Aerospace 73
Parrott, Patrick D. — Solarlite Photovoltaics 122
Paster, Jack — EMC 39
Pastore, Joseph — Sol-Temp Inc. 130
Patel, Chiman R. — Solar Hind Energy Co. 122
Paul, Marguerite M. — Best Energy Systems, Inc. 18

Paul, Terrance — Best Energy Systems, Inc. 18
Pennington, Doug — Current Alternatives 27
Perleberg, Bill 90
Peroni, George — Hydrocap Corp. 60
Peterson, Arnold — Solar Management Corp. 123
Petrella, Ricardo A. — Omega Electronica S.A. 88
Petruzzelli, Tom — Northeast Solar Electric Co. 86
Phillips, Charles — Solar Electric Devices, Inc. 114
Philp, Tom — Specialty Concepts, Inc. 132
Pratt, Richard 94
Price, Stan — Washington Energy Extension Service 153
Provance, Jason — Thick Film Systems 145

Ragsdale, Clyde — Solavolt International 128
Rankin, A. Michael — Solar Electric Systems 116
Reid, Andrew B. — Broken Plow Law Office 20
Reinhardt, Dennis — Cambridge Solar Enterprises, Inc. 21
Richmond, Ronald C. — The Solar Electric Co. Inc. 113
Robbins, Roland W., Jr. — Robbins Engineering, Inc. 102
Roberts, James D. — Energy Management Consultants, Inc. 39
Roberts, Janet — Solarex Corporation 120
Robey, W. P. — GSC Engineered Products 52
Robinson, Rob — Solarwest Electric 127
Rodgers, Clarence — Rodgers & Company Inc. 103
Rosenberg, Paul — Rosenberg Associates 104
Roydhouse, John — Tideland Energy Pty. Ltd. 145
Rydell, Leonard A. 104

Sabisky, Edward — SERI 119
Sah, Chih-Tang — University of Illinois 61
Sams, Arthur D. — Polar Products 93
Samuel, Barbara — Wisconsin Division of State Energy 158
Sandelin, Kenneth D. — Horizon Builders 59
Sanders, John A. — Intersol Power Corporation 64
Schaefer, John F. — New Mexico Solar Energy Institute 84
Schaeffer, J. — Acurex Solar Corporation 2
Schaller, David — Black Hawk Associates, Inc. 18
Schleicher, Carl Solartherm Inc. 126
Schlusser, Larry — Sun Frost 137
Scott, Tom — Energeia 37
Self, George — Chronar Corporation 23
Senghaas, Karl A. — Electrolab Inc. 35

Shimamoto, Edward M. — Delaware Energy Office 29
Shoemaker, Gary L. — Barrett Heating & Air Conditioning Co., Inc. 17
Shupe, John W. — University of Hawaii at Manoa 54
Siegenthaler, John — Appropriate Designs & Construction 9
Simmonds, Donald E. — Lenbrook Industries Limited 70
Simpler, Al — Simpler Solar Systems 108
Skinner, Steve — Solar Electronics International 117
Sklar, Scott — Solar Lobby 123
Sleeper, David — Brook Farm Inc. 20
Slom, M. L. — EASCO Aluminum 34
Slusher, William — Luxtron Inc. 72
Slykhouse, Thomas E. — Slykhouse Engineering 108
Smith, E. E. — Solar Mid-West International 123
Snelson, Greg — Integrated Solar Systems Inc. 64
Soinski, Arthur J. — California Energy Commission 21
Solman, F. J. — MIT Lincoln Lab 74
Somberg, Howard — Global Photovoltaics Specialists Inc. 49
Sotolongo, Tom — AMP Incorporated 8
Spicer, Laurence — Hydrogen Wind, Inc. 60
Starr, Gary — PV Power Corp. 97; Solar Electric Engineering Inc. 114
Starr, Wally — Solar-Eye Products Inc. 121
Stenhouse, Douglas S. — Energy Management Consultants Inc. 39
Stern, Al — ASA Government Marketing Service 13
Stern, Robert L. — Poco Power Corporation 92
Stoll, Bob — GNB Batteries Inc. 50
Strong, Steven J. — Solar Design Associates Inc. 112; Solar Electric Power Company 115
Sullivan, Bruce — Oregon State University 88
Syzmanek, David — REC Specialties, Inc. 99

Tailer, Peter — Windfarm Museum, Inc. 156
Talmage, Peter — Talmage Engineering 142
Tarpey, Ray — Solar Ray 125
Taylor, Robert — EPRI 35
Theobald, Robert 53
Thomas, Michael — Sandia National Labs 105
Thompson, B. — Applied Solar Energy Corp. 9
Thoresen, James — Solar Electric Devices Inc. 114
Throckmorton, Al — Prime Energy Products, Inc. 94
Thurman, Albert — Association of Energy Engineers 14

Timbario, Tom — Mueller Associates, Inc. 81
Toro, J. Enrique — Natural Power, Inc. 82
Torres, Terry — Cosmos Developing Associates, Inc. 27
Trumbull, Dan — ACP Solar Potentials 1

Unger, Earl D. 149
Urso, Mary — Dytek Laboratories, Inc. 32
Usmani, Ishrat H. 150

Vachabach, Ron — Energy Equipment Sales 38
Vanderhoof, Sam — Independent Power Company 62
Verchinski, Steve — Solar Electric Systems (of New Mexico) Inc. 117
Verhalen, Kenneth L. — Kenning 68
Volny, Norbert — Volny Engineering 152

Wagner, J. Bruce — Arizona State University (CSSS) 12
Waizenegger, Jack R. — American Sun 7
Wallace, Vicki — Photron, Inc. 92
Walters, Robert R. — Entech, Inc. 42
Walton, Pat — Solavolt International 128
Warner, Tom — PVI Inc. 96
Webb, W., Jr. — Webb Electronics 154
Wegman, Steven — South Dakota Office of Energy Policy 131
Wiener, Mark — M. Hutton & Co. 60
Welch, Jim — Home Energy Workshop Inc. 57
Whalen, Kathleen — Sunrise Builders 138
White, David C. — MIT Energy Lab 74
White, Jack K. — Photoelectric, Inc. 91
Whitlock, Charles — Airtricity 4
Wilcox, Phil — Natural Sytems, Inc. 83
Wilkins, A. D. Paul — Solar Works 128
Willey, Steve — Backwoods Cabin Electric Systems 15
Williams, Susan — *Energy Review* 40
Wilson, Alex — New England Solar Energy Association 83
Wilson, Will — Sun-Up Solar Security 140
Winters, William W. — Solamerica Corporation 110
Wisman, Rick — Vanner Inc. 151
Wolfe, Tim — Zomeworks Corporation 161

Yudelson, Jerry — Solar Initiative 122

Zahm, Edward Skip — IOTA Engineering 64

Zambas, Peter G.   160
Zatsick, C. H.   160
Zimmermann, Fred — March Mfg. Inc.   73

Geographical Index

ALABAMA

Wyle Laboratories (Huntsville)   159

ALASKA

Popo Agie Inc. (McKinley Park)   94

ARIZONA

University of Arizona (Tucson)   12
Arizona Public Service Co. (Phoenix)   12
Arizona State University (Tempe)   12
Balance of Systems Specialists, Inc. (Scottsdale)   15
DSET Laboratories Inc. (Phoenix)   31
IOTA Engineering Co. (Tucson)   64
New Volt Solar Electric (Tucson)   85
Nordika Systems, Inc. (Scottsdale)   86
Photocomm Incorporated (Scottsdale)   90
Photovoltaics (Scottsdale)   91
Photowatt International, Inc. (Tempe)   92
PVI Publishing, Inc. (Phoenix)   97
RES Photovoltaic Engineering Inc. (Scottsdale)   101
Robbins Engineering, Inc. (Lake Havasu City)   102
Solar Electric Systems (Cave Creek)   116
Solavolt International (Phoenix)   128
SSI (Tempe)   133
Sunamp Systems, Inc. (Scottsdale)   135

ARKANSAS

compuSOLAR (Jasper)   25
Jacuzzi Brothers (Little Rock)   65
Mississippi County Community College (Blythesville)   78

CALIFORNIA

Acro Energy (Ontario)   1
Acurex Solar Corporation (Mountain View)   2
Advanced Energy Corp. (Van Nuys)   2
Airborne Sales (Culver City)   3
Airtricity (Northridge)   4
Alternative Energy Engineering, Inc. (Redway)   5
American Energy Consultants (Canoga Park)   6
American Sun (North Hollywood)   7
Applied Solar Energy Corp. (City of Industry)   9
ARCO Solar, Inc. (Woodland Hills)   10
Berkeley Solar Group (Berkeley)   17
Borrego Solar Systems (Borrego Springs)   19
California Energy Commission (Sacramento)   21
Carbone Investment Management Corporation (Berkeley)   22
Joel Davidson (North Hollywood)   29
The Earth Store (North San Juan)   33
The Earth Store (Nevada City)   33
Electric Power Research Institute (Palo Alto)   35
Enercomp (Davis)   36
Energy Alternatives (Chico)   37
Energy Compliance Systems, Inc. (San Jose)   37
Energy Harvester (San Diego)   38
Energy Management Consultants, Inc. (Los Angeles)   39
Energy Review (Santa Barbara)   40
The Energy Shop, Inc. (Victorville)   41
Ferro Corporation (Santa Barbara)   145
Global Photovoltaic Specialists, Inc. (Woodland Hills)   49
GPL Industries (La Canada)   51
Alvin L. Gregory (Sacramento)   52
Grundfos Pumps Corp. (Clovis)   52
Harbor Freight Salvage (Camarillo)   54
Helionetics (Irvine)   56
Heliotrope General, Inc. (Spring Valley)   56
Hood Miller Associates (San Francisco)   58

Hughes Aircraft (Long Beach)   59
IBE, Inc. (Sun Valley)   61
Independent Energy Producers Association (Sacramento)   62
Independent Home Energy Systems (Yreka)   62
Independent Power Company (Nevada City)   62
Jet Propulsion Labs (Pasadena)   66
Wm. Lamb Co. (North Hollywood)   69
Kyocera International (San Diego)   69
Leveleg (San Diego)   70
MC Solar Engineering (Los Gatos)   75
Lena Menashian (Mountain View)   76
Metal Masters, Inc. (City of Industry)   77
Natural Systems, Inc. (Clearlake)   83
Northcoast Solarworks (Arcata)   86
Northern California Solar Energy Association (Berkeley)   86
Paradise Power Co. (Northridge)   89
Photoelectric, Inc. (San Diego)   91
Photovoltaic Power Systems (Redding)   91
Photron, Inc. (Willits)   92
Poco Power Corporation (San Luis Obispo)   92
Polar Products (Torrance)   93
Polydyne (Menlo Park)   94
PV Network (North Hollywood)   29
PV Power Corp. (Sebastopol)   97
Real Goods Trading Company (Ukiah)   99
Real Goods Trading Company (Willits)   99
Real Goods Trading Company (Santa Rosa)   99
REC Specialties, Inc. (Camarillo)   99
Renewable Power Corp. (Foster City)   101
Sacramento Municipal Utility District (Sacramento)   105
Sandia National Laboratories (Livermore)   105
Scribe International (Santa Ana)   106
SMS Energy Corporation (San Jose)   109
Solar-Audio Visual Electronics (San Francisco)   110
Solar Electrical Systems (Chatsworth)   112
Solar Electric Development Inc. (Irvine)   113
Solar Electric Engineering Inc. (Sebastopol)   114
Solar Electric Manufacturing Co. (Paso Robles)   114
Solar Electric Specialties Co. (Willits)   115
Solar Energies of California (Santee)   118
Solar Energy Industries Association (Woodland Hills)   118

Solar Initiative (Oakland)   122
Solarmode International Corporation (Santa Monica)   123
Solar Sun Batteries (Sacramento)   125
Solarwest Electric (Santa Barbara)   127
Solec International, Inc. (Hawthorne)   129
SolElectric Company (Sacramento)   129
Sollos Inc. (Los Angeles)   130
Sol-Temp Inc. (Capitola)   130
Southern California Edison (Santa Barbara)   131
Specialty Concepts, Inc. (Canoga Park)   132
Spectrolab, Inc. (Sylmar)   132
Stanford University (Stanford)   133
Stern Research Corporation (San Luis Obispo)   134
Strategies Unlimited (Mountain View)   135
Sun-Earth Interface (Palo Alto)   136
Sun Frost (Arcata)   137
Sun Run Company (Mountain View)   139
Sunshine Power Co. (San Jose)   139
Sunwater Construction (Santa Cruz)   140
Tangent Enterprises (Del Mar)   142
TESLAco (Pasadena)   143
Thick Film Systems (Santa Barbara)   145
Frank E. Thompson (Paso Robles)   114
Topaz (San Diego)   146
Trojan Batteries (Santa Fe Springs)   148
United Energy Corp. (Foster City)   149
Varian Associates (Palo Alto)   151
Well-Being Productions (Rough & Ready)   154
Western Solar Products (Santa Ana)   155
Peter G. Zambas (Woodland Hills)   160

COLORADO

A.H.S. Energy Supply (Boulder)   3
Alternologies (Fort Collins)   6
American Solar Energy Society Inc. (Boulder)   7
Black Hawk Associates, Inc. (Denver)   18
Colorado Mountain College (Glenwood Springs)   24
Colorado Technical College (Colorado Springs)   24
Computer Sharing Services, Inc. (Denver)   25
CW Electronic Sales Co. (Denver)   28
Energy Systems Group (Golden)   41

PHOTOVOLTAICS EDITION 179

Gates Energy Products (Denver)   48
Home Energy Workshop (Fort Collins)   57
Insolation Solar (Lakewood)   63
Intersol Power Corporation (Lakewood)   64
Martin Marietta Denver Aerospace (Denver)   73
Now Devices (Denver)   87
Bill Perleberg (Golden)   90
Rockwell International Corporation (Golden)   103
Rocky Mountain Solar Electric (Boulder)   103
Solar Energy Research Institute (Golden)   119
Solar Environmental Engineering (Fort Collins)   120
Solarsoft, Inc. (Snowmass)   125
Solar Utilization News (Estes Park)   126
SunWatt International (Denver)   141
Trinidad State Junior College (Trinidad)   59
Webb Electronics Inc. (Denver)   154
Western Wares (Norwood)   155

CONNECTICUT

Arctic-Kold Energy Products (Bloomfield)   11
Birken Manufacturing Company (Bloomfield)   11
University of Connecticut (Storrs)   26
Oriel Corp. (Stamford)   89
Sunracks (Guilford)   138
Sunsearch (Guilford)   138

DELAWARE

Delaware Department of Administrative Services (Dover)   29
University of Delaware (Wilmington)   29

DISTRICT OF COLUMBIA

American Institute of Architects   6
Biomass Energy Research Association   18
Citizens Energy Project   24
Department of Energy   30
Export Council for Renewable Energy, U.S.   43
National Association of Home Builders   81
National Association of Plumbing, Heating & Cooling Contractors   82
National Association of Solar Contractors   82

Qu.E.S.T. 97
Renewable Energy Institute 100
Rural Electrification Administration 104
Solar Energy Industries Association 118
Solar Energy Institute of North America 118
Solar Lobby 123
Solar Rating and Certification Corp. 124
Worldwatch Institute 159

FLORIDA

ADB Engineers, Inc. (Dania) 2
Advanced Solar Products, Inc. (Coral Gables) 2
Alert 80 Energy Systems (Tavernier) 32
ASA Government Marketing Service (Satellite Beach) 13
Cosmos Developing Associates, Inc. (Vero Beach) 27
Dynamic Solar Products Inc. (Tavernier) 32
Electronic Devices & Controls Inc. (Ft. Lauderdale) 35
Energy Equipment Sales (Lakeland) 38
Energy Systems Leasing (Tavernier) 30
University of Florida (Gainesville) 45
Florida Solar Energy Center (Cape Canaveral) 45
Florida Solar Heating Systems, Inc. (Rockledge) 46
GML Systems, Inc. (Orange Park) 50
Hydrocap Corp. (Miami) 60
Peter T. Landsberg (Gainesville) 69
Tim Merrigan (Cape Canaveral) 77
Pulstar Corporation (Gainesville) 95
Rho Sigma (Hialeah) 102
Simpler Solar Systems (Tallahassee) 108
Solamerica Corporation (Melbourne) 110
Solar America Corporation (Miami) 110
Solar Cells of Florida (Largo) 110
Solar Center (Bradenton) 111
The Solar Connection (Leesburg) 111
Solar Design Consultants, Inc. (Tampa) 112
Solar Designs (Jacksonville) 112
Solar Electric Systems (Pinellas Park) 116
Solar Engineering Company (Cocoa Beach) 119
Solar-Eye Products Inc. (Ft. Lauderdale) 121
Solar One (West Palm Beach) 124
Southeast Residential Experiment Station (Cape Canaveral) 30

PHOTOVOLTAICS EDITION 181

World Efficient Energy Systems (Pinellas Park) 116

GEORGIA

American Society of Heating, Refrigerating & Air Conditioning Engineers (Atlanta) 7
Association of Energy Engineers (Atlanta) 14
Georgia Power Co. (Atlanta) 49
Peachtree Associates, Inc. (Decatur) 90
C. H. Zatsick (Marietta) 160

HAWAII

University of Hawaii at Manoa (Honolulu) 54
University of Hawaii (Honolulu) 55
Hawaiian Solar Electric (Honolulu) 55
Servco International (Honolulu) 107
Servco Pacific Inc. (Honolulu) 107
The Solar Electric Co. Inc. (Honolulu) 113

IDAHO

Backwoods Cabin Electric Systems (Sandpoint) 15

ILLINOIS

Altek Systems, Inc. (Aurora) 5
Bill Becker (Park Forest) 53
Communications Associates (Joliet) 24
Creative Electronics (Chicago) 27
Eltron Research, Inc. (Naperville) 36
Bethe Hagens (Park Forest) 53
University of Illinois (Urbana) 61
Keck & Keck, Architects (Chicago) 67
March Mfg. Inc. (Glenview) 73
K. L. Mathers & Associates (Momence) 74
Methode Electronics (Chicago Ridge) 77
Power Pak (Chicago) 94
Rays Energy Consultants (Springfield) 98
Solar Hind Energy Co. (Carol Stream) 122
Soleq Corporation (Chicago) 130
Tripp-Lite (Chicago) 147

J. C. Whitney (Chicago)   156

INDIANA

Amfridge Corporation (Elkhart)   8
Ball State University (Muncie)   16
Dometic Sales Corp. (Elkhart)   30
Dayton-Walther Corporation (Richmond)   73
Franklin Electric Co., Inc. (Bluffton)   46
Marvel (Richmond)   73
Skyheat Associates (English)   108
SunWatt Corporation (English)   140

IOWA

Hydrogen Wind, Inc. (Lineville)   60
Iowa Energy Policy Council (Des Moines)   65
A. Y. McDonald Mfg. Co. (Dubuque)   75
Solar Mid-West International (New London)   123

KANSAS

Kansas University Solar Energy Club (Lawrence)   67

KENTUCKY

Environmental Alternatives (Louisville)   42

LOUISIANA

Haenni Instruments, Inc. (Kenner)   53

MAINE

Brook Farm Inc. (Falmouth)   20
Maine Office of Energy Resources (Augusta)   72
Rising Sun Enterprises (Bar Harbor)   102
Solar Works Photovoltaics (Port Clyde)   128
Talmage Engineering (Kennebunkport)   142

MARYLAND

Atlantic Solar Power, Inc. (Baltimore)   14
Conservation and Renewable Energy Inquiry and Referral Service (Rockville)   26
Delmarva Power & Light Co. (Denton)   26
Energy Sciences (Rockville)   40
Energy Sciences (Gaithersburg)   40
Integrated Power (Gaithersburg)   63
Monegon, Ltd. (Gaithersburg)   80
Mueller Associates (Baltimore)   81
Semix Inc. (Gaithersburg)   106
Solar Energy Intelligence Report (Silver Spring)   119
Solarex Corporation (Rockville)   120
Solarex Ventures Group (Rockville)   121
Solar International Ltd. (Millersville)   122
Solartherm Inc. (Silver Spring)   126

MASSACHUSETTS

Acheval Wind Electronics (Lowell)   1
American Power Conversion Corporation (Burlington)   6
Bliss Marine (Dedham)   19
Cambridge Photovoltaics Industry (Cambridge)   21
Cambridge Solar Enterprises, Inc. (Cambridge)   21
Crystal Systems, Inc. (Salem)   27
EIC Laboratories (Newton)   35
Energy Materials Corp. (South Lancaster)   39
Energyworks, Inc. (Watertown)   41
Arnold Greene Testing Laboratories, Inc. (Natick)   51
Holec Inc. (North Billerica)   57
Arthur D. Little (Cambridge)   71
Luxtron Inc. (Haverhill)   72
Massachusetts Institute of Technology/Designers Software Exchange (Cambridge)   73
Massachusetts Institute of Technology/Energy Lab (Cambridge)   74
Massachusetts Institute of Technology/Lincoln Lab (Cambridge)   74
Massdesign Architects & Planners, Inc. (Cambridge)   74
Mobil Solar Energy Corporation (Waltham)   79
Northeast Residential Experiment Station (Concord)   30
Polaroid Corporation (Cambridge)   93
PVI Inc. (Boston)   96

Solar Design Associates, Inc. (Lincoln)   112
Solar Electric Power Company (Lincoln)   115
Solar Power Corporation (Woburn)   124
Solenergy Corporation (Woburn)   129
Spire Corporation (Bedford)   132
Sunpower Company (Watertown)   137
Surrette Storage Batteries Co., Inc. (Salem)   141
TriSolar Corporation (Bedford)   147
Warmrays Inc. (Washington)   153
Western New England Solar (Holyoke)   155
Windfarm Museum, Inc. (Vineyard Haven)   156
Xenon Corporation (Wilmington)   159

MICHIGAN

Encon Photovoltaics (Livonia)   36
Enerdyne Solar & Wood Systems (Suttons Bay)   36
Energy Conversion Devices, Inc. (Troy)   38
Institute of Amorphous Studies (Bloomfield Hills)   63
Jordan Energy Institute (Comstock Park)   66
Kalamazoo Energy Office (Kalamazoo)   66
Mechanical Products Inc. (Jackson)   76
Michigan Energy Administration (Lansing)   78
Slykhouse Engineering (Grand Rapids)   108
Solarcon, Inc. (Ann Arbor)   111
Solar Engineering & Contracting (Troy)   120
Solarlite Photovoltaics (Flint)   122
Urban Options (East Lansing)   150

MINNESOTA

Alternative Sources of Energy (Milaca)   5
ATR Electronics, Inc. (St. Paul)   14
Dyna Technology Inc. (Minneapolis)   32, 156
GNB Batteries Inc. (St. Paul)   50
Honeywell Corporation (Bloomington)   58
Minnesota Mining & Manufacturing (St. Paul)   78
Mr. Sun, Inc. (Alexandria)   80
Prime Energy Products, Inc. (Roseville)   94
Red Wing Energy Education Center (Red Wing)   100
Winco (Minneapolis)   156

MISSOURI

Black & Veatch, Engineers – Architects (Kansas City)   18
Kenneth Foster (St. Louis)   46
Londe, Parker, Michels (St. Louis)   71
Missouri Department of Natural Resources (Jefferson City)   79
Washington University (St. Louis)   154

NEBRASKA

Broken Plow Law Office (Chadron)   20
Licor Inc. (Lincoln)   71

NEW HAMPSHIRE

Energy Conservation & Solar Center (Manchester)   37
Drew A. Gillett (Bedford)   49
Hollis Observatory (Nashua)   57
Natural Power, Inc. (New Boston)   82
REV Management Company Inc. (Peterborough)   102
SolarVision, Inc. (Harrisville)   127
Sun Research Inc. (New Durham)   138
Total Environmental Action Inc. (Harrisville)   147

NEW JERSEY

Abacus Controls (Somerville)   1
Chronar Corporation (Princeton)   23
Computer Power Inc. (High Bridge)   25
Danfoss Inc. (Mahwah)   28
EASCO Aluminum (North Brunswick)   33
Gumbs Associates, Inc. (Newark)   53
New Jersey Department of Energy (Newark)   84
Nova Electric Mfg. Co. (Nutley)   87
Princeton Energy Group (Princeton)   95
RCA Corporation (Princeton)   98
Science Associates (Princeton)   106
Solar Energy Corp. (Princeton)   118
Thomas & Betts Corporation (Raritan)   145

NEW MEXICO

Horizon Builders (Raton)   59
Los Alamos National Laboratory (Los Alamos)   72
Microcomputer Design Tools (Albuquerque)   78
New Mexico Solar Energy Institute (Las Cruces)   84
Public Service Company of New Mexico (Albuquerque)   95
Rodgers & Company Inc. (Albuquerque)   103
Solac Builders Ltd. (Albuquerque)   109
Solar Electric Systems (of New Mexico) Inc. (Albuquerque)   117
Solar Works (Santa Fe)   128
Southwest Residential Experiment Station (Las Cruces)   30
WindLight Workshop (Santa Fe)   157
The Workbook (Albuquerque)   158
Zomeworks Corporation (Albuquerque)   161

NEW YORK

Alder/Barbour Marine Systems Inc. (New Rochelle)   4
Aldermaston Sales (Locust Valley)   4
Alternate Energy Transportation Newsletter (NYC)   5
American Society of Mechanical Engineers (NYC)   7
American Standards Testing Bureau Inc. (NYC)   7
Appropriate Designs & Construction (Holland Patent)   9
AREMCO Products, Inc. (Ossining)   11
Barrett Heating & Air Conditioning Co., Inc. (Bayshore)   17
Brookhaven National Laboratory (Upton)   20
Dytek Laboratories Inc. (Bohemia)   32
Electric Vehicles Consultants (NYC)   5
Frost and Sullivan (NYC)   48
Institute of Electrical & Electronic Engineers (NYC)   63
Kayex Corporation (Rochester)   67
Kee Industrial Products, Inc. (Buffalo)   67
Kingston Industries Corp. (Liberty)   68
Long Island Solar Energy Association (Massapequa)   71
New York State Energy Office (Albany)   85
Northeast Solar Electric Co. (Endicott)   86
Paul Rosenberg Associates (Pelham)   104
Silonex Inc. (Plattsburgh)   108
Solartek (West Hurley)   125
Strategic Marketing Inc. (NYC)   134
Sun Arrayed Contracting, Inc. (Roslyn Heights)   136

United Solar Energy Corporation (NYC)   149
Ishrat H. Usmani (NYC)   150
Zipp & Zonen (Bohemia)   160

NORTH CAROLINA

National Climatic Center (Asheville)   82
PNG Conservation (Charlotte)   92
Richard Pratt (Columbus)   94
Wilmore Electronics Co., Inc. (Durham)   156

OHIO

Alpha Solarco (Cincinnati)   4
Appropriate Energy Management (West Jefferson)   9
Battelle Columbus Laboratories (Columbus)   17
CAM-LOK (Cincinnati)   21
Charlins, Inc. (Hudson)   23
Condar Co. (Hiram)   25
Eaton Corporation (Cleveland)   34
Empire Products, Inc. (Cincinnati)   21
Inservco, Inc. (LaGrange)   63
NASA Lewis Research Center (Cleveland)   81
Norcold Inc. (Sidney)   85
Solar Usage Now Inc. (Bascom)   126
Standard Oil of Ohio (SOHIO) (Cleveland)   133
Vanner Inc. (Columbus)   151

OKLAHOMA

GSC Engineered Products (Tulsa)   52
Oklahoma State University (Stillwater)   88
Reading & Bates Development Company (Tulsa)   98

OREGON

ACP Solar Potentials (Salem)   1
Duane's Solar Energy Co. (Salem)   31
Energeia (Eugene)   37
Free Energy Options (Veneta)   47
Lane Energy Center (Cottage Grove)   70
Oregon State University (Eugene)   88
Reading & Bates Development Company (Tulsa)   98

Leonard A. Rydell (Newberg)   104
Solar Ray (Portland)   125
Sunworks (Portland)   141
Tensen Co., Inc. (Portland)   143
Thermal Specialties (Roseburg)   144
Volny Engineering (Bend)   152
Volt Energy Management Service (Portland)   152
Wacker Siltronic Corporation (Portland)   153

PENNSYLVANIA

Ametek, Inc. (Paoli)   8
AMP Incorporated (Harrisburg)   8
Bertoia Studio, Ltd. (Bally)   17
C&D Power Systems (Plymouth Meeting)   22
Drexel University (Philadelphia)   31
Exide Corp. (Horsham)   43
Free Energy Systems Inc. (Lenni)   47
General Electric Company (Philadelphia)   49
Hartell (Ivyland)   54
Kulicke & Soffa Ind. (Horsham)   68
Milton Roy (Ivyland)   54
University of Pennsylvania (Philadelphia)   90
Renewable Energy Projects (Ulster)   101
Solar Electric Devices Inc. (Doylestown)   114
Stoveman Diversified Energy Products (Uwchland)   134
Westinghouse Electric Corporation (Pittsburgh)   156

RHODE ISLAND

Eppley Laboratory (Newport)   43
The Sun Shop (Block Island)   138
Sunrise Technologies, Inc. (Block Island)   138

SOUTH DAKOTA

South Dakota Office of Energy Policy (Pierre)   131

PHOTOVOLTAICS EDITION   189

TENNESSEE

AFG Industries, Inc. (Kingsport)   3
Solar Electronics International (Summertown)   117
Technical Information Center (Oak Ridge)   142
Tennessee Valley Authority (Knoxville)   143

TEXAS

Arkwork Review (Denton)   12
Braden Wire and Metal Products (San Antonio)   20
Carter Wind Systems, Inc. (Burkburnett)   22
Denton County Arkwork (Denton)   12
Dodge Products, Inc. (Houston)   30
Electrolab Inc. (San Antonio)   35
Entech, Inc. (DFW Airport)   41
Environ Energy Systems (Austin)   42
E-Systems (Dallas)   43
M. Hutton & Co. (Dallas)   60
Last Chance Homesteads (El Paso)   70
Photon Energy (El Paso)   91
PV Insiders Report (Dallas)   96
San Patricio Solar (Mathis)   105
Sontek Energy Corporation (Dallas)   131
University of Texas (Austin)   144
Texas Electronics, Inc. (Dallas)   144
Texas Instruments, Inc. (Dallas)   144
Tideland Signal Corp. (Houston)   146
Trinity University (San Antonio)   147
UTL Corporation (Dallas)   150

UTAH

Terralab Engineers (Salt Lake City)   143
Tios Solar Design & Engineering (Salt Lake City)   146

VERMONT

Associates in Rural Development, Inc. (Burlington)   13
Current Alternatives (Northfield)   27
Rick McGowan (Burlington)   76
New England Photoelectric Power Co. (West Brattleboro)   83

New England Solar Energy Association (Brattleboro)   83
North Wind Power Co. (Moretown)   87
Sunnyside Solar (West Brattleboro)   137
Sunrise Builders (Grafton)   138

VIRGINIA

Air Conditioning & Refrigeration Institute (Arlington)   3
American Wind Energy Association (Alexandria)   8
Copley Energy Inc. (Alexandria)   26
Defense Photovoltaics Program Office (Fort Belvoir)   29
Energy Sciences (McLean)   41
Robert B. Haley (Blacksburg)   54
Meridian Corporation (Falls Church)   77
National Technical Information Service (Springfield)   82
PV Energy Systems (Alexandria)   96
Solar Management Corp. (Norfolk)   123
VITA, Inc. (Arlington)   151

WASHINGTON

Boeing Engineering and Construction Company (Seattle)   19
Dynamote Corporation (Seattle)   32
Ecotope, Inc. (Seattle)   34
Edmonds Community College (Lynnwood)   34
Eureka Design (Seattle)   43
John Fluke Manufacturing Co. (Everett)   46
Heart Interface (Federal Way)   55
Homestead Electric (Northport)   58
Integrated Solar Systems Inc. (Anacortes)   64
Seattle University (Seattle)   153
Sun-Up Solar Security (Kent)   140
Trout Creek Power Co. (Republic)   148
Washington Energy Extension Service (Seattle)   153
University of Washington (Seattle)   154

WEST VIRGINIA

Earl D. Unger (Martinsburg)   149

WISCONSIN

Best Energy Systems, Inc. (Necedah)   18
Dale & Associates (Beloit)   28
Mark D. Dostal (Stevens Point)   31
Energy Management Analysis of Madison (Madison)   38
F-Chart Software (Middleton)   44
Flad & Associates (Madison)   44
Globe Battery (Milwaukee)   50
Kenning (Waupaca)   68
Silicon Sensors, Inc. (Dodgeville)   107
Windworks (Mukwonago)   157
Wisconsin Division of State Energy (Madison)   158
University of Wisconsin (Madison)   158

WYOMING

Energy Research & Design Associates (Jackson)   39
Independent Electric Company (Casper)   61

U.S. TERRITORIES
GUAM

Tropical Systems (Mangilao)   148

FOREIGN
ARGENTINA

Omega Electronica S.A. (Buenos Aires)   88

AUSTRALIA

Solarex Pty., Ltd. (Regents Park)   121
Tideland Energy Pty. Ltd. (Brookvale)   145

AUSTRIA

Solar Voltaics (Innsbruck)   127

CANADA

Canadian Solar Industries Association (Ottawa) 22
C&D Batteries (Concord, Ontario) 23
Gold Star Energy Savings Products Ltd. (Saskatoon) 51
Lenbrook Industries Limited (Scarborough, Ontario) 70
Princess Auto (Winnipeg) 95
Solar Cells Ltd. (Burlington, Ontario) 111

ENGLAND

Peter T. Landsberg (Southampton) 69

HONG KONG

Solarex Electric Ltd. 121

JAPAN

ARCO Solar Far East Pty. Ltd., North (Tokyo) 10
Fuji Electric Co. (Yokosuka) 48
Japan Solar Energy Corporation (Kyoto) 65
Komatsu Ltd. (Tokyo) 68
Matsushita Electric Industrial Co. Inc. (Osaka) 75
Mitsubishi Electric Company (Hyogo) 79
Nippon Electric Company, Ltd. (Kanagawa) 85
Sanyo Electric Company, Ltd. (Osaka) 106
Sharp Corporation (Osaka) 107
Toshiba Corporation (Kawasaki) 141

THE NETHERLANDS

Howard Design B.V. (Eindhoven) 59

SINGAPORE

Solar Generators Singapore Pty. Ltd. 121

SWITZERLAND

Solarex S.A. (Gland) 121

WEST GERMANY

Siemans AG (Munich)   107
Suntronic/Solar-Electronic (Hamburg)   139
Telefunken (Heilbronn)   142

Subject Index

actuators 35
aluminum extrusions for PV applications 34, 68
agricultural applications, *See* feeders, gate openers, irrigation, water pumping system design
amorphous silicon cell, R&D 23, 35, 69, 78, 92, 93, 98, 107, 119, 132, 133
anemometry equipment 13, 76
annunciators, manufacture 15
aquacultural applications, R&D 148
architectural design 6, 31, 43, 58, 67, 68, 74, 109, 112, 136, 146, 147, 153
associations, membership 3, 6, 7, 8, 14, 18, 22, 63, 81, 82, 84, 87, 118
audio systems 15, 33, 110
automated equipment design and fabrication studies 50, 69
auxiliary equipment manufacturers, *See* individual equipment listings; DC equipment and suppliers, full-line

backup systems, manufacture 18, 58, 72, 114
balance of systems suppliers, *See* suppliers

ballasts, design and manufacture 64, 99
battery cabling systems, manufacture 5, 77, 123
battery caps, gas recombining hydrolytic 60
battery chargers, manufacture 5, 23, 25, 36, 83, 106, 136
battery manufacturers 23, 43, 48, 50, 61, 77, 125, 141, 148
Also see suppliers
Btu meters, manufacture 26, 102

cadmium/cadmium telluride solar cells, R&D 8, 134
cathodic protection 7, 15, 61, 79, 115, 130, 136
certification/rating 45, 63, 124
circuit breakers, 12-volt DC, manufacture 76
commercialization 10, 55, 65, 69, 74, 105, 120
communications systems, *See* telecommunications
compressor manufacturers 11, 28
computer analysis 9, 12, 15, 16, 18, 19, 39, 44, 77, 98
Also see microcomputer software
concentrating systems 1, 12, 16, 18, 35, 38, 42, 43, 55, 64, 73, 108, 113, 149, 150, 151

195

conductor pastes 145
connectors for PV modules,
  manufacture 8, 21, 145
consulting engineers 2, 7, 9, 10,
  14, 15, 18, 19, 31, 39, 44, 49,
  75, 77, 81, 90, 101, 104, 108,
  109, 112, 122, 130, 133, 140,
  147, 152
controller manufacturers 5, 15,
  36, 56, 87, 91, 102
  Also see suppliers
control systems 35, 36, 37, 58,
  61, 62, 63, 80, 91, 109, 116,
  117, 118, 132, 135, 148
converters, manufacture 1, 56,
  77, 138, 146, 156
copper indium diselenide,
  thin-film cell R&D 30, 31
cuprous sulfide solar cell
  R&D 134
curve tracers, manufacture 96
customized equipment 10, 15,
  21, 79, 92, 115, 140
Czochralski process 8

data acquisition equipment and
  systems 84, 96, 117
DC (12-volt) equipment
  custom unit manufacture 10
  manufacturers 11, 28, 51, 62,
    64, 73, 76, 77, 143
  suppliers 5, 19, 33, 47, 54,
    58, 70, 90, 95, 110, 115, 116,
    126, 134, 155, 156, 157
  Also see lights, pumps,
    refrigerators, etc.
demonstration units for public
  education 42, 45, 108, 139,
  157
  mobile 52, 92, 98
desalination systems 79, 139,
  150

diagnostic systems
  development 84
distributors, *See* suppliers
do-it-yourself kits, *See* how-to

economic analysis 1, 19, 22, 74,
  75, 145, 152
education programs, college
  level 16, 24, 34, 45, 55, 57,
  59, 62, 66, 71, 75, 78, 86, 89,
  100, 117, 134, 136, 137, 143,
  147, 153, 160
  Also see information
  dissemination; seminars;
  workshops
electric vehicles 5, 29, 53, 113
energy audits 80, 98, 101
  Also see consulting
  engineering; computer analysis
energy farms development 4, 8,
  149
energy libraries 42, 66, 67, 72,
  74
energy management, *See*
  consulting engineering;
  computer analysis
evaporative coolers,
  PV-powered 68
export activities 43, 57, 86, 88,
  93, 99, 115, 141

fabrication process studies, solar
  cells, *See* solar cell R&D
fans, 12-volt DC,
  manufacture 28, 68
F-chart 25, 44, 111
feasibility studies 2, 39, 49, 77,
  88
feeders, PV-powered 20, 35
film-fed growth, solar cell
  R&D 79, 107
flowmeters, manufacture 102

frequency changers,
    manufacture 87
Fresnel lens systems 16, 42, 43,
    79, 149

gallium arsenide solar cells 79,
    132, 150, 151
gate openers, PV-powered 20,
    35
gen-sets 32, 115, 148, 153
glass manufacture for PV use 3
glow discharge deposition, solar
    cells 78

how-to
    information/designs/kits 5, 13,
        39, 54, 80, 94, 124, 126, 154,
        157
    Also see workshops
hybrid system design 10, 13, 17,
    42, 53, 84, 87, 93, 99, 101,
    108, 117, 131, 139, 140
hydrogen fuel production 60,
    81, 145
hydro systems 5, 15, 33, 62, 93,
    125
    Also see water pumping
    systems

indium phosphide, solar cell
    R&D 134
industrial applications of PV, *See*
    large-scale PV applications
information dissemination
    journals/magazines 5, 13, 40,
        53, 84, 87, 91, 96, 120, 123,
        126, 127, 158
    newsletters/fact sheets 5, 10,
        22, 29, 31, 40, 45, 55, 84, 96,
        97, 116, 119, 128
    slide programs 25, 71, 72

information referral agencies,
    governmental 29, 45, 55, 65,
        66, 78, 79, 82, 84, 85, 100,
        104, 118, 119, 131, 142, 158
information referral services 24,
    26, 42, 67, 76, 89, 128, 138,
    150, 157, 160
    Also see demonstration units;
    energy libraries; seminars;
    workshops
instrumentation, manufacture 30,
    35, 43, 53, 57, 63, 71, 76, 82,
    102, 106, 117, 132, 144
integrators, manufacture 57, 71,
    82, 106
international applications and
    studies 10, 53, 74, 120, 141
    third world projects 13, 19,
        21, 73, 93, 108, 120, 124,
        128, 147, 148, 150, 151
    Also see export; licensing
inverters, manufacture 1, 2, 18,
    25, 27, 32, 33, 55, 56, 87, 91,
    99, 143, 146, 147, 151, 156
investment analysis and
    consulting 22, 59, 96, 97,
    102, 118, 122, 135, 160
irrigation systems 75, 79, 93,
    110, 139
    Also see water pumping
    systems

kits, *See* how-to

large-scale PV system design 2,
    4, 10, 12, 18, 19, 27, 29, 39,
    42, 44, 59, 64, 74, 78, 81, 87,
    101, 112, 120, 124, 128, 141,
    149
    Also see utility-scale
licensing, overseas 21, 141, 152

lights, 12-volt DC,
   manufacture   54, 58, 64, 99
   low-pressure sodium   99
line-tie/interconnect systems, See
   utility-interactive systems

mail-order suppliers   4, 5, 15, 17,
   19, 24, 29, 31, 33, 36, 40, 41,
   54, 62, 69, 80, 99, 110, 116,
   117, 121, 126, 127, 128, 140,
   148, 14, 156, 157
manufacturers, PV
   cells/modules/systems, See
   photovoltaics
manufacturing process
   development   11, 35, 49, 120
marine systems,
   design/manufacture   7, 11, 47,
   86, 109, 115, 126, 136, 141,
   148
   specialty equipment
   suppliers   19, 21, 32, 122,
   128, 130, 138
market assessment and
   research   13, 21, 22, 48, 59,
   74, 77, 80, 88, 90, 98, 101,
   120, 122, 134, 135, 159, 160
metering devices/meterboards,
   manufacture   15, 26, 30, 43,
   46, 53, 57, 60, 82, 128
microcomputer software for PV
   design applications   6, 17, 25,
   34, 36, 38, 41, 44, 71, 54, 78,
   90, 95, 104, 106, 111, 118,
   120, 125, 138, 139, 140, 152,
   155, 158
   software library   74
   Also see computer analysis
mobile demonstration units   52,
   92, 98
mobile power systems   18, 60,
   64

module assembly machines   69,
   70, 120, 124, 132
modules, See photovoltaics
mounts, See support structures
multimeters, manufacture   46

navigational aids/systems   7, 10,
   117, 130, 146, 147
   Also see marine systems
novelties, manufacture and
   supply   4, 94, 114, 126, 136

packaged systems, See suppliers
Pelton power unit
   manufacture   15
performance prediction,
   semiconductor materials   35
performance testing, See testing
photoelectric control cells   107,
   118
photosensors, R&D   9
photovoltaic breeders   81, 149
photovoltaic cell manufacturing
   equipment   11, 67, 132
photovoltaic cell/module/system
   manufacturers   4, 6, 10, 15,
   19, 23, 27, 31, 35, 38, 42, 48,
   50, 64, 73, 85, 90, 91, 96, 98,
   106, 107, 108, 111, 114, 120,
   121, 124, 128, 129, 131, 132,
   133, 140, 142, 145, 149
   Also see suppliers, full-line
photovoltaic cell, R&D   10, 12,
   16, 17, 26, 29, 45, 53, 55, 56,
   61, 66, 85, 106, 108, 120,
   129, 133, 140, 144, 145, 154
   amorphous silicon   23, 35, 69,
   78, 92, 93, 98, 107, 119, 132,
   133
   cadmium/cadmium telluride,
   thin-film   8, 134
   copper indium diselenide,

thin-film 30, 31
  cuprous sulfide 134
  dendritic web process 156
  film-fed growth 79, 107
  gallium arsenide 79, 132, 150, 151
  glow-discharge deposition 78
  indium phosphide 134
  metallization process 130
  nonvacuum alternatives to solar cell fabrication 35
  performance prediction of semiconductor materials 35
  polycrystalline, thin-film 19, 34, 94, 109, 119, 132, 153
  research sponsorship 30, 66, 119
  ribbon silicon 35, 39, 65, 71, 79, 128, 146, 156
  screen-printed process 75
  semicrystalline 120, 153
  silicon-on-ceramic 58
  zinc phosphide 134
photovoltaic module assembly equipment manufacture 69, 70, 120, 124
photovoltaic "shingles" 23, 49
photovoltaic/solar thermal system design 3, 52, 78, 79, 109, 131, 140, 150
photovoltaic systems R&D 6, 10, 20, 61, 69, 81, 103, 105, 108, 109, 119, 120, 140
photovoltaic window systems 130, 140
publications, *See* information dissemination
pump manufacturers 11, 15, 23, 51, 52, 54, 65, 68, 73, 75, 93, 110, 136, 161
  submersible 46, 52, 89
  Also see water pumping systems
pyranometers, manufacture 30, 43, 57, 71, 82, 102, 144, 160
pyroheliometers, manufacture 30, 43, 160

raceway interconnect system 145
radiometers, manufacture 30, 43, 57, 60
recombiners 23, 60
recreational vehicle (RV) applications 7, 32, 64, 68, 83, 85, 105, 109, 116, 128, 130, 136
refrigeration systems design/R&D 3, 7, 13, 19, 62, 72, 76, 83, 86, 93, 128, 130
refrigerator (DC) manufacturers 4, 8, 11, 28, 30, 73, 85, 86, 93, 126, 137
remote system design 7, 9, 10, 15, 25, 29, 57, 61, 62, 70, 79, 83, 85, 86, 105, 109, 113, 115, 116, 127, 128
  Also see international applications, third world projects; rural electrification
research sponsorship 30, 66, 119
ribbon process development 35, 39, 65, 71, 79, 128, 146, 156
rural electrification 13, 19, 57, 73, 104, 130, 150
  Also see international applications, third world projects

safety analysis, *See* testing
seminar services 15, 16, 36, 55, 62, 71, 75, 76, 79, 84, 92, 97, 108, 122, 123, 136, 148

silicon cells, *See* photovoltaic cells
sizing, *See* consulting engineers; suppliers
solar cells, *See* photovoltaic cells
solar simulators, R&D/manufacture   89, 102, 132, 159
space applications   9, 132
strip chart recorders, manufacture   83, 102, 106
sub-arrays, manufacture   124
suppliers, full-line/manufacturers' representatives, retailers (system sizing and design capabilities)   1, 2, 5, 7, 9, 10, 14, 15, 17, 20, 21, 25, 27, 28, 31, 33, 36, 37, 38, 40, 41, 43, 46, 47, 50, 51, 52, 55, 57, 58, 59, 60, 62, 63, 64, 68, 70, 76, 83, 85, 86, 88, 92, 94, 97, 99, 101, 103, 105, 109, 110, 111, 113, 114, 116, 117, 118, 119, 121, 122, 123, 124, 125, 126, 127, 128, 129, 134, 137, 140, 142, 144, 147, 155, 157
Also see mail-order suppliers
support structures   8, 10, 34, 58, 62, 68, 71, 81, 105, 120, 123, 129
surplus suppliers   3, 46, 54, 95
solar cells   46, 108

telecommunications systems   7, 10, 21, 47, 57, 60, 62, 70, 79, 86, 109, 115, 117, 120, 124, 128, 130, 145, 147
temperature sensors, manufacture   102, 106, 144
testing equipment, manufacture   96, 102, 109, 123

testing laboratories   7, 9, 30, 31, 45, 51, 66, 71, 74, 81, 84, 92, 105, 120, 124, 129, 143, 154, 159
testing programs   13, 18, 21, 22, 63, 101, 109, 123, 126, 142, 152, 155, 159
thermometers, manufacture   26, 53
thermostats, differential, manufacture   102
thick-film conductor pastes   145
third world PV projects, *See* international applications
tools, 12-volt DC, manufacture   3, 143
track homes   113
tracking devices/systems   41, 45, 64, 79, 89, 102, 117, 123, 131, 149, 161
transportation systems   7, 10, 75
Also see electric vehicles

uninterruptible power systems (UPS)   1, 25, 50, 72, 87
utility-interactive systems design   1, 6, 9, 26, 45, 84, 90, 93, 95, 97, 101, 105, 109, 112, 113, 114, 117, 119, 128, 136
utility-scale power plant design   1, 2, 10, 12, 18, 26, 29, 49, 64, 73, 81, 95, 101, 105, 120, 124, 131

village power systems, *See* international applications, third world projects; rural electrification
voltage regulators, manufacture   15, 56, 82, 87, 102, 124, 128, 132, 135, 146

waste heat recovery
   systems   103, 109, 131
water pumping system design   3,
   7, 9, 10, 11, 13, 15, 21, 23,
   35, 40, 51, 52, 61, 62, 65, 68,
   75, 76, 79, 83, 84, 86, 90,
   103, 110, 114, 115, 116, 117,
   119, 123, 124, 126, 127, 128,
   130, 141, 146, 147, 150
   Also see pump manufacturers

water purification systems   79,
   139, 100, 150
wind farm development   4, 8
wind/PV systems   10, 17, 22, 67,
   77, 84, 87, 93, 103, 109, 127,
   139, 148, 150, 156, 157
workshops   15, 16, 39, 42, 45,
   53, 55, 57, 62, 66, 67, 71, 72,
   76, 79, 84, 108, 122, 128,
   137, 148
   Also see how-to; seminars